Contracts and Liability

for Builders and Remodelers

Fourth Edition

David S. Jaffe

Department of Legal Services
Regulatory and Legal Affairs Division
National Association of Home Builders

Home Builder Press®
National Association of Home Builders
1201 15th Street, NW
Washington, DC 20005-2800
(800) 223-2665

Contracts and Liability for Builders and Remodelers

Fourth Edition

ISBN 0-86718-410-8

Cover by David Rhodes, Art Director, Home Builder Press

Printed in the United States of America

Library of Congress Cataloging-in-Publication Data

Contracts and liability for builders and remodelers / David S. Jaffe, Department of Legal Services, Regulatory and Legal Affairs Division, National Association of Home Builders. — 4th ed.

p. cm.

Includes bibliographical references.

ISBN 0-86718-410-8

1. Construction contracts — United States.

I. National Association of Home Builders (U.S.).

KF902.Z9C695 1996
346.73′078624—dc20
[347.30678624] 95-48241
CIP

For further information, please contact:

Home Builder Press®
National Association of Home Builders
15th and M Streets, NW
Washington, DC 20005-2800
(800) 223-2665

About the Author

David S. Jaffe, is Staff Counsel in the Department of Legal Services of the NAHB Regulatory and Legal Affairs Division. He regularly advises builders and remodeler members on contract and liability issues as well as other aspects of construction law. He drafted the previous edition of *Contracts and Liabilities for Builders and Remodelers* and is a regular speaker at NAHB educational programs on contract and liability issues.

1/96 Rand/Kirby 5M

Contents

Figures

Acknowledgments

Special thanks go to the following organizations and people:

- Home Builders Association of Louisville and the Greater San Antonio Builders Association for their permission to reprint portions of their model contracts.
- American Law Institute and Committee on Continuing Professional Education, American Bar Association, and Neil Schemm, Esquire, for their permission to reprint from "Subcontract Forms From the Subcontractor's Perspective," published in the *Practical Real Estate Lawyer*, September 1986.
- Thomas J. Stipanowich, Attorney, Lexington, Kentucky, and editor of the *Construction Lawyer* for permission to quote from "Legal Exposure of the Design/Build Participants: The View of the General Contractor," Vol 15, No. 3 (August 1995), pp. 8-28.
- Ross S. Robbins, Vice President, Bainbridge, Inc., Englewood, Colorado, for assistance with the discussion of agreements between builders and realtors in Chapter 9.
- Ron Burton, Director, NAHB Energy and Home Environment Department; Rhonda Daniels, NAHB Office of Regulatory Counsel; and Phoebe Schlanger, NAHB Environmental Counsel for assistance with Chapter 5.
- Dan Johnson, Senior Technical Advisor, NAHB Technology and Codes Department, for ideas for the sample residential construction agreement in the Appendix.

Reviewers

M. M. (Mike) Weiss, Jr., President of Weiss & Co., Carmel, Indiana, and Leonard A. White, Esq., of Goldstein, Handler, and White, Bethesda, Maryland, deserved special appreciation for reviewing the entire manuscript as well as the outline. Special thanks go to the following people who reviewed the outline and/or various chapters or parts of the manuscript: Wilson Barnes, AIA, Coordinator, Broward Programs, Department of Construction Management, College of Engineering and Design, Davie, Florida; Robert Bell, CGR, President, Bell Remodeling, Inc., Duluth, Minnesota; Bob Blayden, CGR, President, Blayden Design/Build, Renton, Washington; Tom Cooper, Associate Professor, Building Science Department, Auburn University, Alabama; John Geoffroy, President, CDCI, Atlanta Georgia; Richard Graf, President, R. W. Graf, Inc., Knoxville, Tennessee; Robert Hall, President, R. H. Building Contractors, Inc., Los Gatos, California; Jerry Householder, Chairman, Department of Construction, Louisiana State University, Baton Rouge, Louisiana; Robert A. Kelzer, CGR, President, R. Systems, St. Petersburg, Florida; Charlie Knutson, CGR, President, Knutson Brothers Builders, Inc, West Allis, Wisconsin; Michael McCalvy, CGR, President, Remodeling Consultants, Inc., Minneapolis, Minnesota; Bob Merz, Arbitrator, Construction Arbitration Associates, Ltd., Roswell, Georgia; Cynthia Milloy, CGR, General Manager, Oaken Hammer Design and Remodeling, Annapolis, Maryland; Tim Mrozowkski, Professor, Michigan State University, East Lansing, Michigan; Kenneth E. Ormond, Jr., Esq., Columbia, South Carolina; Philip Lee Russell, President, Energy Smart Corporation, Pensacola, Florida; Tony Thompson, CGR, President, Remodeling Services Unlimited, Columbia, South Carolina; Joedy Sharpe, President, Sharpe Construction Co., Inc., Danville, Kentucky; and Carol Smith, Owner, *Home Address*, Littleton, Colorado.

NAHB and Senior Staff

This fourth edition of *Contracts and Liability for Builders and Remodelers* was produced under the general direction of Kent Colton, NAHB Executive Vice President/CEO in association with NAHB staff members Jim Delizia, Staff Vice President, Member and Association Relations; Adrienne Ash, Assistant Staff Vice President, Publishing and Information Services; Rosanne O'Connor, Director of Publications; Doris M. Tennyson, Director, Special Projects/Senior Editor and Project Editor; David Rhodes, Art Director; and Carolyn Kamara, Editorial Assistant.

Introduction

During the past few years the members of the National Association of Home Builders (NAHB) have become more and more concerned about the many lawsuits filed and the increasingly large judgments assessed against home builders, remodelers, and others in the construction industry. This fourth edition of *Contracts and Liability for Builders and Remodelers* responds to these concerns of NAHB members.

NAHB's goal in producing this book is to help builders, remodelers, and general contractors control or limit liability in the construction and sales processes by (a) acquainting them with legal principles and laws they are likely to encounter in their businesses on a day-to-day basis and (b) helping them write more complete contracts. The legal problems builders, remodelers, and others face derive primarily from two sources: (a) vague or nonexistent contract documents and (b) unspoken and unmet expectations. This book explains how legal disputes with buyers, owners, and other trades often arise when contracts are vague and expectations remain unspoken and unmet. It also illustrates ways of minimizing these disputes. Subcontractors, suppliers, and other industry professionals can also benefit from much of the advice in this book.

This book is written for anyone in the residential building and remodeling industries: a builder building a first custom home, an experienced builder moving into a new market, an experienced remodeler negotiating a design-build project, or a subcontractor who is presented with a contract by a builder or a remodeler.

This edition of *Contracts and Liability* includes summaries of actual cases decided by courts so that the readers can see how the legal concepts discussed in the book are applied in typical construction cases. Readers will also find new commentary and contract provisions have been added throughout the book. Two useful sample contracts may be found in the appendix at the end of the book.

Warning—The forms and procedures contained in this book are simply illustrations. They should not be used without the approval of an attorney experienced in construction contract law. The suggested contract provisions do not and cannot apply to every situation. Applicable law differs widely among the states, and in many cases, local municipal law may be applicable. Some of the provisions may not even apply to a particular construction project, and in such a case, they might weaken a builder's or remodeler's position with respect to that contract.

This information is intended only to identify certain issues and highlight some alternatives. It is offered as a service to builders and remodelers to help them identify potential problems that they need to consider carefully and discuss with an attorney.

In the sample language the words that appear in brackets [ ] are included to suggest to the reader a way or ways the provision could be modified.

Builders and remodelers should have their attorneys prepare documents that meet their particular needs.

One • Reducing Liability

This chapter and this book discuss ways in which builders and remodelers can minimize the liability that might arise out of their construction and sales processes. For purposes of this discussion, the term *liability* means legal responsibility, either civil or criminal, as determined by a judge or a jury. In noncriminal matters, liability may be determined by an arbitrator or some other neutral third party as part of an alternative dispute resolution procedure (ADR) procedure. But liability is not limited to those instances in which builders or remodelers are defending an action (see Chapter 2). It does include situations in which builders or remodelers initiate proceedings, such as when they attempt to recover final payment. Because each proceeding costs money, builders and remodelers have an interest in reducing both types of liability.

In addition liability includes unanticipated costs associated with or resulting from disputes with owners and purchasers, even if the matters in dispute were not actually litigated or arbitrated. This type of liability includes the cost of performing additional work without pay because an item was not adequately addressed in the contract. For example, a builder constructing a custom home on the owner's lot may not be entitled to recover the added cost of removing railroad ties discovered during excavation if the contract is silent regarding changed or differing site conditions and if the builder's estimate does not provide for such a contingency (see Chapter 2).

A well-written contract is a critical tool for reducing liability, and most of this book is devoted to the words, phrases, and concepts contributing to that end. Using a well-written contract to reduce liability, however, does not mean using a contract that is one-sided in favor of the builder or remodeler. For example, builders and remodelers should not attempt to shift to purchasers or owners certain risks that are more properly the builder's or remodeler's responsibility. Similarly builders and remodelers should not attempt to absolve themselves of all liability. Rather, they should strive to create a contract that—

- is a product of the parties' negotiations
- describes in detail the rights and obligations of the parties
- fairly allocates the risks inherent in the project

Such a contract provides the parties with a mutual understanding of the contract terms and conditions, thereby improving the chances that each party's expectations will be met. And if each party's expectations are met, they are less likely to end up in litigation or arbitration.

This blueprint for reducing liability recognizes that some disputes result when the builder's or remodeler's work fails to substantially comply with the plans and specifications. Some disputes result from the owner's or purchaser's inexcusable material failure to comply with the payment terms. But often disputes between the parties involve good faith issues that their contracts do not adequately address, either because the parties did not consider these issues or because they failed to write them into the contract.

Thus, a well-written contract helps prevent disputes from arising during construction because the parties have settled potentially troublesome matters such as the scope of the work, the materials to be used, and the time and method of payment before the project begins. In addition, if the contract contains a limited warranty and disclaims all other expressed and implied warranties (in those states that allow it), the contract also helps prevent disputes after construction is completed.

Another potential source of liability is work performed by third parties such as subcontractors. A well-written contract fairly allocates risk to other major parties in the construction process (such as subcontractors, architects, and engineers) by obtaining contractual indemnities from them. (*Indemnity* means shifting financial loss from the one required to pay it to the party that caused the loss. Indemnity rests on the premise that one party has been obligated to pay money that in justice another person ought to pay. A contract can shift the risk of loss from one party to another.) Thus, where a builder or remodeler is liable to a purchaser or an owner for work performed by a subcontractor, the builder or remodeler may be entitled to recover from the subcontractor if the subcontract contains an indemnity clause (see Chapter 8).

Although builders and remodelers can reduce their potential liability in a number of ways throughout the construction process and afterwards, they will find no magic steps they can take to guarantee that they will not be sued or end up in arbitration or mediation. The two most important methods of reducing liability are to make a strong commitment to customer service and to use well-written contracts. These and other actions that have successfully reduced liability for builders and remodelers are listed in Figure 1-1. Builders and remodelers should keep these preventive measures in mind at all times.

Finally, although a well-written contract requires communication between the parties, it is no substitute for ongoing communication with the owner or buyer during the project. If a problem arises during construction or thereafter, builders or remodelers should not be content to sit back, smug in the knowledge that in a legal battle they would prevail under the contract. Even if builders or remodelers win such a law suit, they could still lose money and time, to say nothing of the possible negative publicity such a battle would generate.

The warranty, like the contract, should clearly express the intent of the parties. The limited warranty describes the problems for which the builder or remodeler will be responsible after completion of the project, and it limits the time after completion within which the owner may complain about the work. Moreover, if a builder or remodeler warrants workmanship and materials in a warranty, he or she must provide some standards against which to judge whether he or she is complying. In the absence of such standards, the builder or remodeler and the owner may find themselves at odds if the owner's personal standards differ from industry standards. If the parties do not spell out the construction standards, they run the risk of having a court decide for them.

The best existing source is *Quality Standards for the Professional Remodeler* (second edition).[1] No comparable book of official standards exists for new construction, but builders also might refer to this book for applicable standards. Accordingly the contract might include a statement such as one that follows.

> All workmanship shall conform to the standards found in the publication *Quality Standards for the Professional Remodeler* (second edition), and if an item is not covered in that publication, industry standards shall govern.

Today home builders and remodelers may be liable to an owner or a purchaser under several legal theories of recovery, and because each theory has met with some success in various cases, a typical construction complaint often alleges several of these causes of action. Some of the popular grounds for law suits brought by homeowners, particularly against builders or remodelers, include—

- breach of contract
- breach of express warranty
- breach of implied warranty
- negligence
- fraud and misrepresentation
- strict liability

Breach of Contract

The owner is entitled to receive the product he or she contracted for, and the builder or remodeler cannot unilaterally provide a lower-quality product or refuse to perform in accordance with the contract terms. Failure to carry out an obligation of the contract may constitute a breach of contract.

Illustrative Case

Kaiser contracted with Fishman for the construction of a beach-front dwelling. Fishman knowingly deviated from the plans and specifications provided by Kaiser. The original plans called for 20 pilings, but Fishman placed only 14 pilings, and 6 of them were drilled in incorrect locations. The court determined that the resulting dwelling was inferior to that called for by the plans and specifications. The court concluded that Fishman deliberately failed to comply with the contract terms, so it awarded Kaiser "cost to cure" damages, which would give him with the house he bargained for, rather than the repair damages argued for by Fishman.[2]

Figure 1-1. Actions to Prevent Liability Problems

- Use a well-written contract for all transactions in the sales and construction process. All parties to the contract should read and understand all provisions before signing the document.
- Make a strong commitment to customer service. Such a commitment involves conscientious customer relations, quick follow-up on punchlists and callbacks, and correcting the problem promptly in some cases (even when the remodeler or builder is not at fault) because that course probably will prove less expensive than litigating the matter later.
- Regularly examine your insurance program to make sure it is comprehensive. Fill in any gaps in insurance coverage.
- Develop realistic construction schedules that allow enough time to do the work well, for any interruptions that might be caused by the anticipated weather, and for minor unexpected problems with deliveries.
- Emphasize safety in all aspects of the business. Insist on safe procedures and use of safety equipment.
- Follow all building codes and other regulations to the letter. Avoid cutting corners.
- Use reputable subcontractors, design professionals, and real estate brokers; establish sound working relationships with them and all parties in the construction process; promote a cooperative attitude; and avoid polarizing the parties to all contracts.
- During the sales and construction process, make only promises you can keep and keep those you make. Review brochures, advertising, and other marketing devices for unintended promises or warranties.
- Treat prospects', owners', and buyers' concerns seriously.
- Obtain signed change orders (showing the price) for any alterations to the original construction plans, specifications, or other aspects of the job.
- Comply with all laws prohibiting discrimination based on race, religion, creed, color, national origin, sex, marital or family status, age, arrest record, or disability. Review employment policies, practices, and materials as well as brochures, advertising, and other marketing devices for unintended discrimination.
- Use indemnity clauses in contracts with design professionals and subcontractors.
- Disclaim and/or limit the duration of implied warranties or obtain waivers of such warranties in states where waivers are permissible.
- Consider alternative dispute resolution techniques, such as arbitration or mediation, in lieu of litigation.
- Operate a building or remodeling business in an ethical, legal, and carefully organized way with an emphasis on customer service to greatly reduce your exposure to liability.
- Retain an attorney knowledgeable about construction for review of contracts and periodic review of operations. However in the event of a dispute builders and remodelers can seek the assistance of some local builders associations or Remodelors™ Councils. Some local affiliates of the National Association of Home Builders (NAHB) have committees to handle complaints from home buyers and owners. NAHB encourages its local associations and councils to establish complaint-handling systems. For more information on establishing such a program, builders and remodelers can contact the NAHB Consumer Affairs Department.[3]

Breach of Express Warranty

Express warranties are any statements (oral or written) that the work will meet certain standards. Breach of warranty occurs when such a statement proves to be false.

Illustrative Case

A prospective purchaser, who wanted to build a home with a view of a valley, asked a designer to look at a lot to determine its suitability as a homesite. The designer advised the prospective purchaser that the lot was suitable, that the home would fit properly on the lot, and that the house would have a view. However the house could not be placed in the position chosen by the purchaser because of zoning restrictions and restrictive covenants.

The court held that in assuring the purchaser that the lot selected was suitable for the purpose of the contract and that the house could be placed on it properly the designer warranted that certain facts existed, which did not, and that by making such a warranty the designer promised to pay damages if the facts were not as warranted. Accordingly the purchaser was entitled to compensation for those injuries which the designer had reason to foresee as a probable result of its breach.[4]

Breach of Implied Warranty

Implied warranties are warranties imposed on builders by state courts and statutes. They exist without the builder making any oral or written promise, which distinguishes them from express warranties. The courts in many states have held that builders of a house warrant (by implication) that the house is habitable and that it was built in a workmanlike manner. Because the law imposes this warranty, it is assumed to be a part of the contract even if it is not stated in the contract. An implied warranty covers both materials and workmanship and remains in effect for a reasonable time. What is reasonable will vary from case to case. However the warranty may last as long as 5 or 10 years after substantial completion of the project.

Some states also subject remodeling contracts to an implied warranty.

Illustrative Case

The Gambles purchased a lot and entered into a contract with Main for the construction of a home on the lot. Before purchasing the lot, the Gambles had sought Main's advice as to the suitability of the lot, and Main had advised them that it was suitable for a homesite. Less than a year after they moved into the home, the Gambles noticed a slip developing in their front yard and attendant seepage from the absorption field of their septic system. The Gambles sued Main. They asked the court to instruct the jury that, by building on the site Main gave "a warranty or guaranty that the site was fit for those purposes." But the court refused and the Gambles appealed.

The West Virginia Supreme Court acknowledged that a majority of states extend some form of protection to the purchaser of a new home, and that this protection is usually an implied warranty of habitability or fitness that requires builders to construct the dwelling in a workmanlike manner and to deliver the property reasonably fit for its intended use of human habitation. The court also confirmed that such a warranty existed in West Virginia. However the court concluded that the implied warranty of habitability or fitness does not extend to soil conditions that the builder is unaware of or could not have discovered by the exercise of reasonable care.[5]

Fraud and Misrepresentation

The elements necessary to establish an action for fraud and misrepresentation include—

- a false representation of a material (important) fact
- knowledge or belief on the part of the speaker (or writer) that the representation is false or that he or she does not have sufficient information to make it
- an intent to induce the other person to act or refrain from acting based upon the misrepresentation
- justifiable reliance upon the representation on the part of the injured party
- damage to the plaintiff resulting from such reliance

Under certain circumstances builders and remodelers have a duty to disclose facts regarding the condition of the property, such as a hidden defect that the homeowner could not discover upon reasonable inspection of the property. In such a case silence may constitute fraud or misrepresentation.

Illustrative Case

Home buyers sought damages from a builder-developer because the new homes that they bought in Vorhees Township, New Jersey, were constructed near a hazardous waste dump site known as the Buzby Landfill. The families alleged that, although the developers were aware of the existence and hazards of the landfill, they did not disclose those facts to the families when they bought their homes.

The court observed that in the case of on-site conditions, courts have imposed affirmative obligations on sellers to disclose information materially affecting the value of property, and "there is no logical reason why a certain class of sellers and brokers should not disclose offsite matters that materially affect the value of the property."

Thus, in New Jersey a builder-developer of residential real estate or a broker representing the builder may be liable for not disclosing off-site physical conditions that they knew about and that were not known nor readily observable by the buyer. However the conditions must be of sufficient materiality to affect the habitability, use, or enjoyment of the property, so as to render the property substantially less desirable or valuable to the objectively reasonable buyer.[6]

Negligence

Negligence is the failure to use such care as a reasonably prudent and careful person would use under the same or similar circumstances. In negligence cases certain conduct is imposed by the law rather than by the will or intention of the parties. Negligence deals with the law of torts. (A tort action seeks to remedy a civil wrong. It is based on public policy, whereas a contract action is based on the mutual promises of the parties). Negligence is the breach of an individual's duty to exercise reasonable care so as not to cause unreasonable risk of harm to another.

Generally the elements necessary to establish an action for negligence are—

- a legal duty owed by one person to another to exercise the degree of care demanded by the circumstances to protect the other person from foreseeable harm
- a breach of that duty
- a direct causal connection between the breach and the resulting injury
- actual loss or damage

Owners may assert negligence suits together with or instead of breach of contract actions because they might recover certain damages in a negligence suit that they cannot recover in a breach of contract action. Courts have found negligence when a construction defect resulted in personal injury, damage to other property, and loss in value of the property.

Illustrative Case

ABC Builders, Inc., built a house at the toe of a hillside in 1968. In 1978 heavy rains caused the hillside to slide down and push against the house. The slide destroyed all but the top floor of the house. The Phillipses, the third owners of the house, sued ABC on both warranty and negligence theories. The court held that ABC had a duty to furnish a safe location for a residential structure and that the duty did not depend on whether damage might arise within the confines of the lot boundaries or from forces originating beyond its limits. ABC was liable to the Phillipses for negligence in selecting the site because the evidence established that ABC was an experienced builder and in possession of extensive knowledge about the location and that it should have known of the apparent dangers of the hillside as a building site.[7]

Strict Liability

A person engaged in the business of manufacturing or selling products for use or consumption is strictly liable where the following actions occur: the person places the product on the market, knowing that it is to be used without inspection for defects, and a defective or dangerous condition of the product causes personal injuries, death, or property damage to foreseeable users or consumers, or even to a mere bystander.

Strict liability requires producers of injurious products to be liable for injuries caused by those products, whether or not the producers are negligent. It differs from other liability claims, which are based on some kind of fault: either negligence or breach of contract. In one line of cases the courts have maintained that (a) strict liability should apply to "mass producers of homes," on the grounds that a home is a product like other manufactured goods, and (b) no meaningful distinction exists between builders of mass-produced homes and manufacturers and or sellers of other products.

Illustrative Case

Kriegler filed an action for physical damage sustained as a result of the failure of a radiant heating system in a home constructed by Eichler Homes, Inc. Kriegler purchased the home in 1957 from the Resings, who purchased it from Eichler in 1952. In November 1959 the radiant heating system failed. Emergency and final repairs required removal and storage of furniture and forced Kriegler and his family to acquire other shelter temporarily. The court held that Eichler was liable to Kriegler on the basis of strict liability. After acknowledging that strict liability had been applied in the state only to manufacturers, retailers, and suppliers of personal property and rejected as to sales of real estate, the court concluded that no meaningful distinctions existed between Eichler's mass production and sale of homes and the sale of automobiles.[8]

Uniform Commercial Code

For the reasons discussed below, builders and remodelers should be familiar with the Uniform Commercial Code (UCC). The UCC is a uniform body of law that governs a wide variety of business transactions, including the sale of goods. The UCC does not apply to the sale of real estate nor to transactions where the predominant factor is the provision of services. Typically, construction contracts involve the performance of services; nonetheless, builders, remodelers, and other participants in the building industry should be familiar with the UCC. Should materials or supplies fail to perform, the UCC contains specific warranties to be implied in sales contracts. Just as the builder impliedly warrants that his or her work is habitable and performed in a workmanlike manner, the seller of goods impliedly warrants to the buyer that his or her goods are merchantable (UCC §2-314). And an implied warranty of fitness for a particular purpose arises when the seller knows that the buyer is relying on the seller to furnish goods suitable for a particular purpose. Breach of these warranties by the seller may entitle the buyer to recover incidental and consequential damages.

Illustrative Case

A roofing subcontractor sought to recover damages resulting from its reroofing of a college. The roofer's cause of action was based in part on breach of warranty in connection with the roofing materials originally sold to it by the manufacturers. The court held that the roofing materials supplied to the roofing subcontractor were not fit for the particular use required by the roofing subcontractor and that the roofing company was entitled to recover the cost of reroofing and the interest it paid on the funds it borrowed to replace its working capital consumed in the reroofing.[9]

Two • Contract Between Builder and Buyer or Owner

A written contract records the exact terms of an agreement between parties. The contract defines the scope of the work and the price of the product and allocates the risks inherent in a particular transaction between the parties. This chapter of *Contracts and Liability for Builders and Remodelers* should help builders and their attorneys write construction contracts that lessen the builders' risks.

Many provisions in ready-made contract forms, such as the American Institute of Architects' Document A201, expose builders to considerable liability during the construction process. No boilerplate contract clause is suitable for every situation. Therefore builders should critically examine every transaction into which they enter to foresee contingencies or events that could impair the transaction's benefits to them. They should make sure that each of their contracts is written to provide them with adequate protection from events that may expose them to liability.

Contracts allocate the responsibilities and the risks associated with those responsibilities. Some risks are standard and cannot be assigned or transferred to others. For example, the builder bears the responsibility for and the risks associated with ensuring that a home is built according to plans and specifications. The allocation of other risks and responsibilities is not always so clear. For example, who should buy the builder's risk insurance? Not necessarily the builder.

Builders should not shy away from risks as long as they get paid for taking those risks. The object is to avoid as many risks as possible for the same price.

This chapter examines the purpose of various clauses that might be included in a construction contract and presents sample language that may be used in construction or sales contracts between buyers (or owners) and builders. The explanation for each provision helps readers understand the sample language.

These suggested provisions do not address every contingency, and they do not apply to all construction and sales agreements. Neither does this chapter cover all of the provisions that builders and buyers should consider for a sales or construction contract. For example, parties should negotiate and document who will be responsible for miscellaneous items such as—

- Payment of hazard insurance and utility fees during construction
- Proration of tax and insurance costs upon sale of the property
- Bond requirements, if any
- Payment of impact fees, if any

These and other provisions, if written correctly, may further protect builders against unnecessary liability.

Builders who offer an insured warranty to their customers should redraft or not use those warranty provisions contained in this book that conflict with their particular insured warranty program.

Contracts are legal documents that greatly determine a builder's liability if a home buyer alleges that the builder failed to perform the contract obligations. An attorney experienced in construction contract law should therefore prepare (or at least review) any such documents before a builder signs them. To find such an attorney the builder could contact the local home builders association or the local branches of the Associated General Contractors or of the Associated Builders and Contractors.

The sample language that appears in brackets presents an option for the reader. Depending on the particular transaction the reader may choose one or the other, both, or neither of the possible wordings presented.

Warning—The National Association of Home Builders has provided this guide and sample contract language merely to point out the types of provisions of which builders should be aware. These suggested provisions should not be used unless they have been reviewed by an attorney.

This chapter addresses the major sources of potential liability that a builder may face in two different types of construction contracts: (a) a contract with a buyer who owns the lot or land on which the house will be built and (b) a contract to construct a house on a lot owned by the builder under which the house and the lot will be sold together.

The parts of the contract include the agreement, the general conditions, and the plans and specifications. Other documents or clauses may be included in a contract.

The Agreement

The agreement is the document that the parties sign. All the other documents that make up the contract such as the general conditions and the plans and specifications are named in the agreement as being part of the contract.

The builder's warranties and warranty limitations often also are drafted separately (see Express and Implied Warranties and Limited Warranty in Chapter 6). If the agreement specifically incorporates these other documents, their requirements and limitations will be a part of the agreement. The agreement should specifically and accurately describe the referenced documents by date, number of pages, plan number, name, and any other appropriate information. The requirements contained in the incorporated documents should not conflict with each other nor with the terms of the agreement. Examples of contract documents appear in Figure 2-1.

Common sense will dictate which of the following provisions apply to a particular builder's needs.

Caption

The caption is the heading or introductory part of a legal document. It should include the names of all of the home buyers (both husband and wife, for example, if they are buying jointly). The full legal name of each party should appear in all documents in the transaction. If a purchaser is a corporation or other business entity, the documents should include the full name and type of business entity. The names of the parties should be exactly the same in the contract as in all other project documents. The caption should also give the location and a legal description of the property (for example, lot, block, and subdivision).

> This contract specifies the terms between _(buyer's name)_ (the customer) and _(builder's name)_ (the builder) to construct the _(home or other structure)_ on the property located at _(legal description of property)_.

The Contract Documents

As discussed earlier, a typical written residential construction contract consists of several different documents, and these are incorporated in the contract by reference in the agreement. For example, the plans and specifications are usually incorporated in the contract by reference as a matter of convenience. Generally, when a written contract incorporates other documents by reference, a single contract is created which includes the contents of the incorporated documents. To avoid any confusion about which plans and specifications are incorporated, the agreement should specifically and accurately describe the referenced documents by date, number of pages, plan number, name, and any other appropriate information. In addition the parties should initial each page of the documents at the time of signing the contract. They also should initial the drawing(s) and every page of the specifications to avoid (a) an accusation that a different page has been substituted or (b) a plea of ignorance if a disagreement arises between the owner and the builder. Moreover, the prudent builder will include an "order of preference" clause in the contract. This clause specifies whether the plans or the specifications take precedence in case of a conflict between the two. The parties are free to decide the order of precedence when the builder provides the specifications but not the plans. He or she may want the specifications to take precedence in case of a conflict.

> The terms of this contract include the conditions of this contract and, by reference, the provisions in the other documents specifically listed below. The terms of this agreement prevail over any conflicting provisions in the documents incorporated by reference. In the case of a difference, discrepancy between, or ambiguity in, the plans and specifications, the parties agree that the specifications shall govern.

Figure 2-1. Examples of Contract Documents

- The agreement includes all the items that can vary from project to project such as the names and addresses of the parties to the contract, the name and location of the project, the dates for beginning and substantial completion of the project, a list of the other documents included in the contract, a list of the work involved in the contract documents but to be done by others, price of the contract, payments and how they are to be made, and signatures. The agreement includes any part of the contract in which blanks must be filled in.
- Additions are changes to the contract (issued before execution of the agreement).
- General conditions set forth those conditions that do not vary from one contract to another, such as settling disputes, grounds for terminating or suspending the contract, and procedures for making changes in the work. They spell out responsibilities not assigned by other parts of the contract (such as insurance and bonds). They also cover ownership of the drawings, specifications, and other documents.
- Specifications describe the work to be done, the material to be used, and how it will be used or installed.
- Drawings (plans) are pictorial representations of the project.
- Special conditions are any modifications or additions to the general conditions.
- A warranty is a statement (oral or in writing) that the work will meet certain standards and a promise by the builder to correct deficiencies in workmanship and materials.
- Change orders are modifications to the contract (issued after execution of the agreement).
- A draw schedule describes when payment is due; it is often tied to commencement or completion of certain stages of the work.
- An allowance schedule identifies those items of work that are not sufficiently detailed in the contract documents to enable the parties to determine the actual cost of the item.

The builder shall perform all of the work that is required by this agreement and any documents incorporated by reference below.

Except for written modifications executed by both parties subsequent to the execution of this contract (for example, change orders), the terms of the contract are limited to the provisions contained in this agreement and the other documents described as follows:

- (title of document), dated the ____________ day of ______________________, 19____ consisting of ____ pages;
- (title of document), dated the ____________ day of ______________________, 19____ consisting of ____ pages;
- (title of document), dated the ____________ day of ______________________, 19____ consisting of ____ pages;

Illustrative Case

In this action by the builders to recover the reasonable value of labor and materials furnished in the construction of a house and garage, an issue before the court was whether the contract required the parties to arbitrate the matter before pursuing litigation. The written building contract provided that all work contemplated within the terms of the agreement was more fully described in the architect's plans and specifications. The specifications provided that general conditions in certain pages of a specified form issued by the American Institute of Architects should govern. On this basis the court was warranted in finding that articles 39 and 40 of the AIA form, which included a provision for arbitration, were included within the agreement of the parties.[10]

Ownership of Plans and Specifications

If the buyer provides the plans, the builder should specifically impose liability for defects in the plans on the buyer. Builders who provide the plans for a house, however, should expressly prohibit buyers from giving or selling those plans to other potential buyers or builders. The best way to ensure enforcement of this latter provision is to copyright the plans through the Copyright Office of the Library of Congress. Similarly, if the buyer provides the plans, the builder should make sure that the buyer owns the plans because, if the buyer does not own the plans, the builder may be sued for copyright infringement.

Register of Copyrights
Library of Congress
Washington, DC 20559

Sample language is provided below for both the buyer's plans and for the builder's plans.

> **Buyer's Plans**—The builder agrees to construct the home in accordance with the building plans supplied by the buyer _(reference by title, author, and number of pages)_. The builder assumes no responsibility or liability for defects in the design or engineering of these plans. The buyer warrants that the plans and specifications are adequate and that the builder can rely on them. The buyer will be liable for any damages caused by defective building plans and specifications, including but not limited to additional costs caused by delay in substantial completion of the home, additional costs for materials necessitated by any changes, and additional labor costs, pro rata overhead, and profit on the additional work.
>
> The buyer represents that (a) the house plans provided to the builder are the result of an original design, (b) the buyer is the sole owner of the design, and (c) the buyer has exclusive rights, including copyright, in and to the design as represented in the structure, plans, specifications, and drawings for the house. The buyer agrees to indemnify and hold the builder harmless from all claims of third parties for copyright infringement or conversion which may be asserted as a result of the builder's construction of the house for the buyer.
>
> **Builder's Plans**—The builder has provided the building plans and specifications to be used under this contract _(reference by title, date, author, and number of pages)_. The buyer has no ownership rights in the plans and specifications used under this contract, and the buyer will be liable to the builder for the amount of lost profits and all consequential damages for the reuse or resale of these copyrighted plans.
>
> The builder makes no representations or warranties about the quality of these plans except those specifically provided in the limited warranty references in this contract.

Time of Commencement and Substantial Completion

The date the work begins should be no earlier than the date of execution of the contract. Instead of a specific date, a flexible starting date may be established by using a notice to proceed; for example, "The work shall commence on the date stipulated in the notice to proceed." The builder should have evidence of the buyer's financing before commencing work (see Price, Deposit, and Payment later in this chapter). Before building on a lot owned by the buyer, the builder should require evidence of a valid title and a copy of the current survey, deed restrictions, liens, and easements, if any. These items may help protect the builder's lien rights if a lien must be filed on the property for lack of payment, or they may identify potential site problems that the builder can avoid.

The contract completion date should be expressed as a number of days (preferably calendar days) from the contract date or notice to proceed and should be based upon substantial completion of the work rather than final or full completion. The contract should clearly define what constitutes substantial completion, for example, the date of the final inspection or the date of the certificate of occupancy or temporary certificate.

One legal definition of substantial completion is when the buyer can use the project for its intended purpose. A finding that the work is substantially complete entitles the builder to the contract price, although the owner is entitled to an offset for the cost of completing or correcting the deficient work.

However substantial completion should not be pegged to the buyer's satisfaction, a subjective standard. If construction is connected to an interstate land sale, the builder must comply with time periods for delivery under the Interstate Land Sales Full Disclosure Act, 15 United States Code sec. 1701 et seq. (1982) (see 24 Code of Federal Regulations parts 1700-1730).

The contract should include a provision for extending the time of completion, as is done in Liquidated Damages later in this chapter.

> Assuming all conditions are satisfied and weather permits, the work to be performed under this contract shall be substantially completed no later than _(number)_ calendar days from _(date)_.

Illustrative Cases

Case 1—A contract for the sale and erection of a prefabricated house provided for completion of the house on September 27, 1961. When the purchasers arrived on the scene at 4 p.m. on the specified date, at the builder's behest for a final inspection, the workers were still "slapping on siding, laying floors, bulldozing the yard, hooking up utilities, and so on," and the purchasers refused to accept the house, despite the foreman's assurances that the home would be completed by 5:30 p.m. The court held that the builder substantially complied with the contract on the date specified when only service walk, some grading, and blacktopping were left undone at 5:30 p.m. The court found that no substantial sum was required to complete the items left undone, and the purchasers could have resided in the home at that time.[11]

Case 2—Aleda Construction Co., Inc., sued the Winns claiming that the Winns breached a written construction contract by failing to pay the balance due for the construction of a house. The contract provided that the Winns would make payments at various specified stages of completion and that the funds would be paid "in consideration of the covenants and agreements being strictly performed . . ." by Aleda. The contract provided for final payment when the house was "fully complete and [Aleda] supplied [the Winns] with a final survey and executed Release of Liens." At trial Aleda's president conceded that (a) several items were incomplete (which he estimated would cost $500), and (b) Aleda had not furnished the Winns with a final survey or a release of liens. Finding for the Winns, the court concluded that Aleda had not "strictly" performed, and that the house was not "fully" completed.[12]

Price, Deposit, and Payment

This section of the contract specifies and defines the builder's compensation for the construction of the project. It also instructs the parties in the manner and time of payment. To achieve these goals, this section should—

- Include the cost of the work. Some builders charge buyers a total fixed price for the project. Others bill on a cost-plus-fee or cost-plus-percentage basis. The cost-plus contract should specify the percentage or fee required, the guaranteed maximum price (if any), what constitutes costs (see section on costs), how the fixed fee is to be adjusted for change orders, the type of fee schedule, and all anticipated costs and fees, when applicable. In either case, the contract should specify (in both words and numbers) (a) the total price of the construction or sales transaction and (b) the amount of the buyer's deposit to be paid to the builder upon execution of the contract.
- Identify the first and, if applicable, the second mortgage or deed of trust.
- Specify that work covered by change order fall within the general scope of the work contemplated by the contract, and specify a percentage of profit for additional work. Change orders for deductions from the work usually involve no reduction of profit, and this fact should be clearly stated.
- Establish the amount of the deposit and the due dates for all payments. The builder should check applicable state and local laws regarding escrow accounts. (A sample state law appears in Figure 2-2.) Parties often arbitrarily set due dates for payments. The due dates should be mutually acceptable to builder and buyer. They often reflect specific stages of construction. Builders can specify that payments come due at the start of a construction activity rather than at the end (for example, a payment might come due upon the beginning of framing rather than at the completion of the foundation). Buyers rarely dispute when an activity begins but sometimes question when one ends. The contract must account for the time necessary for the builder to prepare an application for payment and for the buyer to make the payment. For instance, the contract might state, "The owner has ten (10) days from the statement date to pay the builder."
- Require interest for late payments. For example, "Payments due under the contract but not paid shall incur daily interest at the rate of ______ percent (____%) from the date payment is due." Instead of charging interest on late payments, the contract might provide for a late payment fee similar to liquidated damages. Such a late fee might be included in a cost-plus-fee contract as well as in a fixed-cost contract. An alternate clause might read as follows:

Figure 2-2. Sample State Law Regarding Escrow Account

Pursuant to Maryland statutory law (Md. Real Property Code Ann. §10-506 [1988]), a custom home builder must include in each custom home contract an escrow account requirement notice. The escrow account notice shall be on a separate page of the custom home contract; and the escrow account requirement notice required shall state:

Escrow Account Requirement

Unless your contract is financed by a mortgage issued by a federally-chartered financial institution or a financial institution supervised under the Financial Institutions Article of the Annotated Code of Maryland, or unless all deposits, escrow money, binder money, or any other money paid in advance, or is paid to the licensed broker, to be held in the escrow account of the broker, Maryland law requires that all consideration exceeding 5 percent of the total contract price which is paid by a buyer to a custom home builder in advance of completion of the labor, or receipt of the materials for which the consideration is paid shall be deposited in an escrow account and paid out of that account only for certain purposes specified by law. To ensure this, the law requires that your builder may only accept such payment in the name of the escrow account. Thus, you should make out your check to "(name of builder), escrow account." Records of payments out of this account must be carefully maintained by your builder, and the builder must permit you reasonable access to escrow account records. Your builder, however, may choose to establish a separate account for your project which will require your signature for any withdrawals."

Buyer hereby acknowledges that late payment by the buyer to the builder of progress payments and final payment may cause the builder to incur costs not contemplated elsewhere in this contract, the exact amount of which will be difficult to ascertain. Accordingly any sum due the builder under this contract shall be paid within ten (10) days of written request. If full payment is not received within ten (10) days, the buyer shall pay ten percent (10%) of the total amount overdue as a late charge. The parties agree that such late charge represents a fair and reasonable estimate of the costs the builder will incur by reason of late payment by the buyer.

- Specify whether retainage is allowed. If the contract allows retainage, it should specify when it will be disbursed.
- Require evidence of financing. The builder should be allowed to verify that the buyer has satisfactory financing, including an allowance for subsequent change orders. The buyer can be made to produce evidence of financing before the builder is obligated to start work.
- Satisfy any special state or local requirements. Some states require that the contract specifically identify certain expenses (for example, Washington state requires that the sales tax be listed separately from the total price).

Fixed-Price Contract—The buyer agrees to buy and the builder agrees to ___[construct and/or sell]___ the building ___[and lot]___ for the consideration of _______ dollars ($_________), the total price. The buyer will pay __________ dollars ($_________) to the builder as a deposit [if one is required] upon signing this contract. The buyer will make the following interim payments of ________________ dollars ($________):

___(List payments and tie them to a construction activity such as the beginning of drywall.)___

The buyer will make a final payment in the amount of ______________ dollars ($_________) to the builder at final settlement ___[within (   ) days of substantial completion]___.

Cost-Plus-Fee Contract—The buyer agrees to buy and the builder agrees to ___[construct and/or sell]___ the project for the consideration of the builder's actual costs and expenses (as defined in Section ______ of this contract, Costs to Be Reimbursed) plus a fixed fee of __________ dollars ($_________). The buyers will pay __________ dollars ($_____) as a deposit upon signing this contract ___[or on or before some date other than the date the contract is signed]___ and will make a final payment (including all costs and expenses plus the fee specified above) to the builder at final settlement ___[or within (   ) days of substantial completion)___. Substantial completion has occurred when the buyer can use the project for its intended purpose.

If the buyer requires an addition to the scope of

the work, the builder shall account for the cost of these additions separately and shall be reimbursed _______ percent (______%) of the cost of the changes.

Costs to Be Reimbursed

This section is applicable to cost-plus-percentage or cost-plus-fee contracts. The list below suggests costs that may need to be reimbursed in a construction project, but it may not include all costs to be reimbursed on every job. The builder should include all relevant items and especially every contingency negotiated with the buyer. Builders should check this section against their general ledgers and other accounting records to make sure all appropriate costs are included.

Many contracts will also include a section regarding costs not to be reimbursed. These provisions should be carefully reviewed because, by definition, they exclude some of the builder's actual costs. These items would be paid out of the percentage or fee added to the defined costs and therefore would cut into the builder's profit.

Because of the numerous costs associated with building a home, disputes are likely to occur over costs. Therefore the parties should take great care in identifying which costs are reimbursable and which are not. Moreover the actual costs incurred by the builder must be reasonable. Builders should be prepared to show that (a) any money they claim to have paid out for materials and work on the job were necessary, and (b) they deducted costs incurred because of the inefficiency or incompetence of their agents or employees.

The term *cost of work* means costs necessarily incurred in the builder's proper performance of the work contained in the contract documents, including the following items:

- Wages, benefits, and costs of contributions and assessments for workers' compensation, unemployment insurance, Social Security, taxes, or any other costs incurred by the builder for labor during performance of this contract.
- Salary, wage, or hourly rate for supervision by builder. The rate for the builder's time should be agreed upon by the parties.
- Salaries for builder's field personnel to the extent their time is spent on work required by this contract. This payment of salaries includes (but is not limited to) work performed at shops, on the construction site, or in transporting materials or equipment except where such work is not in accordance with the plans and specifications.
- For out-of-town work the builder's and the builder's agents' reasonable travel and subsistence expenses incurred performing this contract.
- Payments made by the builder to subcontractors for work performed according to subcontracts under this agreement.
- Cost of land and all land development expenditures associated with or apportioned to this project.
- Cost of all materials, supplies, and equipment incorporated in the work and consumed while doing the work; cost less salvage value of such items used but not consumed that remain the property of the builder, including all costs of installing, repairing and replacing, removing, transporting, and delivering the machinery and equipment except in cases in which the builder is in error.
- Rental charges consistent with those prevailing in the area for machinery and equipment used at the construction site, whether rented from the builder or others, including all costs of installing, repairing and replacing, removing, transporting, and delivering the machinery and equipment.
- Small tools (any tool that costs less than ____________ dollars [$________] and consumables are not to be included in the cost.
- Sales, use, excise, or any other taxes related to the work imposed by governmental authorities.
- Impact fees, exactions, royalties, charges, inspection costs, or any other fees related to the work imposed by governmental authorities.
- Permit fees, royalties, or damages for infringement of patents and costs of defending related lawsuits for items specified by the buyer, deposits lost for causes other than the builder's negligence, and tap fees.
- Expenses for telephone service at the site, telephone calls, telegrams, delivery fees, and other similar petty cash items related to work.
- Costs incurred for safety and security at the job site.
- Costs incurred for building code and regulatory compliance.
- All landscaping and backfilling necessary under the contract documents, including the cost of all grading, removing or planting trees, removing snow, frost breaking, water pumping, excavating

and related work, delivering the necessary materials to the site, and hauling away excess fill material or trash and debris.

- Builder's risk or other insurance, soil fees and civil engineering fees, performance bonds, and labor and supplier's bonds in an amount equal to 100 percent (100%) of the maximum costs.
- Costs associated with differing site conditions, as provided for in Section ____ of this contract.

Draw Schedule and Application for Progress Payment

The builder and the buyer may negotiate any form of payment schedule that is mutually agreeable. In the absence of a payment schedule, the builder is not entitled to receive payment until the completed structure is delivered to and accepted by the owner. Typically, where the owner is financing the project, the method of payment is through "progress payments" from the owner to the builder. One alternative allows the builder to establish a draw schedule with the home buyer and construction lender. With a draw schedule a specified amount (for example, 10 percent) is sometimes paid up front. The builder can withdraw from the account at successive phases during construction, and the lender may hold a retainage until the structure is accepted by the buyer.

A draw schedule should require the buyer to make or authorize each payment to the builder within a specified number of days after the builder's application for payment. The builder uses the application for payment document, along with various supporting documents, to establish the builder's right to payment. Before requesting final payment, the builder should ensure that all claims have been settled and should obtain signed confirmations from the buyer that, based on the final inspection, the contract requirements have been fulfilled. Alternatively the contract may provide for inspection by the lender or another third party at certain stages of the construction, with payment due directly from the lender upon satisfactory inspection.

The parties should stipulate specifically who will schedule the inspections and who will pay the inspector. In any case the contract should specify whether the progress payments provision includes payment not only for the work in place but also for materials and equipment suitably stored at the site or at an off-site location agreed upon by the owner.[13]

This provision might give the owner the right to withhold progress payments otherwise due to the builder under certain circumstances enumerated in the contract, for example, if the builder fails to pay his or her subcontractors or suppliers. Similarly this provision might state that (a) the builder can stop work until he or she is paid, and (b) if payment is not made within so many days after it is due, the builder can terminate the contract.

The following sample language provides for two methods of payment.

> **Regular Periodic Payment**—The builder shall prepare [weekly, biweekly, or monthly] itemized statements for the buyer that specify all costs incurred by the builder in furtherance of performance of the terms of this contract. The buyer will place adequate funds in an escrow account from which payments are made for the cost of the work. Each draw will be based on an application for payment submitted to the buyer by the builder. The buyer will make funds available to the builder within __________ ( ) days of receipt of the statement (less __________ percent ________ [%] retainage) as provided for in the Notices provision of this contract.
>
> **Payments Tied to Construction Activity**—The builder shall prepare [weekly, biweekly, or monthly] itemized statements for the buyer that specify all costs incurred by the builder in furtherance of performance of the terms of this contract.
>
> The buyer will place adequate funds in an escrow account from which payments are made for the cost of the work. Each draw will be based on beginning various phases of the work as described below. The buyer will make funds available to the builder within __________ ( ) days of receipt of the notice (less _____ percent ______ (%) retainage) as provided for in the Notices provision of this contract.
>
> The buyer agrees to make progress payments as follows:
>
> _____ % upon signing the contract
> _____ % upon beginning of the slab (foundation)
> _____ % upon beginning of framing
> _____ % upon beginning of drywall
>
> The buyer shall make final payment to the builder, including the entire unpaid balance of the cost of the work and all other compensation due to the builder, as defined in Sections ____, Price and Deposit, and ______, Costs to Be Reimbursed, of

this contract at final settlement [or within ______ ( ) days after substantial completion of the work].

Liquidated Damages and Unavoidable Delay

A liquidated damages clause provides for the payment of a predetermined amount of money in the event of a breach of contract, such as the inexcusable late completion of the work. In this context the word *liquidated*, means determined or settled and refers to damages that are agreed to in advance.[14] Thus liquidated damages are amounts of money agreed to ahead of time that the builder will pay to the buyer if the builder does not complete the contract by the contract completion date. Similarly liquidated damages are amounts of money agreed to ahead of time that the builder may retain if the buyer does not complete the contract. This type of provision may minimize damages in the event of a lawsuit brought on the basis of a default.

The liquidated damages clause should not impose a penalty on either party, but it should reasonably approximate the actual losses one party will incur if the other party does not perform. Estimating actual losses a buyer has suffered can be difficult, and a liquidated damages clause eliminates this problem. A primary purpose of the clause is to compensate buyers for expenses, such as hotel costs, until they can occupy their homes. A court that determines the liquidated damages clause is, in fact, a penalty may void the provision. The nonbreaching party then may sue the other party for all actual and consequential damages actually suffered.

> If the buyer incurs additional costs as a consequence of the builder's failure to perform the terms of this contract by its stated completion date (for example, housing, moving, or travel expenses), the builder will compensate the buyer in the amount of ________________ dollars ($________) for each calendar day that the project is delayed beyond substantial completion (when the buyer can use the project for its intended purpose [or otherwise defined]). If no actual damages occur, the builder does not pay any.
>
> The builder will use its best efforts to complete construction of the project before the completion date. However, if reasons beyond the builder's control cause an unavoidable delay in the progress of construction (including, but not limited to, such factors as the unavailability of materials, inclement weather, strikes, changes in governmental regulation, acts of governmental agencies or their employees, acts of God, or the failure of the architect or the buyer to perform their responsibilities under this contract), the builder may request an extension of the date of completion of the project in writing. The request must be made within 30 days of the beginning of the cause of the delay. The buyer shall not recover any compensation from the builder for delay caused by differing site conditions, as defined and provided for in Section ____, Differing Site Conditions, of this contract.
>
> This liquidated damages remedy is exclusive of all other legal and equitable remedies.
>
> Should the buyer fail to carry out this Agreement, at the option of the builder, all moneys paid hereon shall be forfeited to the builder as liquidated damages unless the buyer is unable to secure financing as set forth in paragraph ____ of this contract.

Illustrative Case

A husband and wife who breached a contract for the construction of a custom home sought the return of a $5,000 deposit that they had given to the contractor. The contract provided that the deposit was to be forfeited as liquidated damages if the couple breached the contract. The couple argued that the provision was a penalty and therefore unenforceable. Finding for the builder, the court concluded that the amount set as liquidated damages represented a reasonable approximation of the contractor's actual damages and was not a penalty because (a) the deposit amounted to less than 4 percent of the total anticipated price, (b) the contractor kept the lot off the market from 90 to 120 days, and (c) the contractor had lost $12,500 net profit provided for in the contract.[15]

Bonus for Early Completion

As a corollary to the liquidated damages clause, builders often include a bonus provision that entitles the builder to a per diem sum of money for finishing early. The contract should address the question of whether an extension of the completion date also extends the bonus date. Some contracts address this issue as shown in the sample language below. Others may simply state: "Under no circumstances will the bonus date be extended." Builders generally should opt for the language that appears below.

> If the builder substantially completes the work before the substantial completion date in Section ____ of this contract, the buyer will pay the builder the sum of _____________ dollars ($_________) for each calendar day between the date when the work is substantially completed and the substantial completion date set by the contract. Unavoidable delays as described above will extend the bonus date.

Notices and Records

To take advantage of certain rights provided for in the contract, each party often must give certain notice to the other party. The contract may provide the form of the notice and the time within which the notice must be given. Failure to follow these notice requirements may jeopardize a claim. For example, the contract might provide that the builder is entitled to notice of defective work within so many days of the discovery of the defective work. It might also give the builder several days to correct the work before the owner may declare the builder to be in default of the contract. If the owner fails to provide the builder with such notice, the owner may be precluded from terminating the builder.

Similarly the contract might provide that if the builder discovers a hidden subsurface condition, the builder must give written notice of the condition to the owner within so many days of the discovery. And if the builder proceeds with the work without first notifying the owner, the owner may balk at paying the builder for the additional work, because the builder precluded the owner from choosing whether to proceed with the work.

The builder should keep all records related to the project for at least the duration of the project. Ideally the builder should store the records until all applicable statutes of limitations have expired. The statute of limitations is the period of time within which a buyer must file a lawsuit. That period of time can run for many years, and under certain circumstances the courts may extend the limits indefinitely.

> Any notice required or permitted to be delivered under this contract should be mailed to the parties at the following addresses:

_____________________	_____________________
(builder)	(buyer)
_____________________	_____________________
(street address)	(street address)
_____________________	_____________________
(city, state, zip)	(city, state, zip)

> For purposes of this contract, notice is received when sent by certified mail, postage prepaid, return receipt requested via the United States Postal Service.

Effective Date and Signatures

Ideally all parties to the contract sign the document at the same time. If that is not possible, however, this provision assures that the builder is not bound by the terms of the contract until the builder (or an authorized representative) has signed the contract.

If any changes or additions are made to the contract after it is printed or typed, both the builder and the buyer or buyers must initial each revision for it to be valid. Of course a change order should cover any changes to the work after the contract is signed in accordance with the requirements of the section on Changes in Work. Any changes or additions to the contract after it is signed require a written amendment according to the Entire Agreement and Severability section presented later in this chapter.

The buyer and builder should sign at least three original contract documents. Many builders use hard copies generated on their computers. The parties should execute the contract in their proper capacities, as individual buyers or sellers or as partners, officers, or directors of a partnership or corporation that is a party to the contract. The parties should use their complete legal names on all contract documents and any subsequent change orders or contract amendments. Once both parties have signed the document, they should complete and initial the date-of-contract provision.

> This contract has no force or effect and will not be binding upon the builder until it is accepted and executed by the builder. This contract shall become effective on the date designated below or the day it is executed by both parties, whichever is later.
>
> We, the undersigned, have read and understand and agree to each of the provisions of this contract and hereby acknowledge receipt of a copy of this contract.

______________________ (buyer's signature)

______________________ (name of builder, corporate name where applicable)

By ______________________ (authorized signature)

Date ______________________ Title ______________________

______________________ Date ______________________

This contract is dated, and becomes effective:

______________________ (month, day, year)

______________________ (buyer's initials) ______________________ (builder's initials)

Illustrative Case

In an action for breach of a construction contract, Grubb claimed that a corporation of which he was a principal stockholder—Double Cousins Company—should have been the defendant in a lawsuit and not Grubb personally. Grubb and Fitzgerald entered into a contract with Cloven which opened as follows:

> To Double Cousins Company, Agent 1.
>
> The undersigned, herein called the Buyer, offers to buy, subject to the terms set forth herein, the following property:

Then follows a blank space in which had been written by hand:

> House on Buyer's lot according to Plan 5392-R of Architect's York and Schenke of New York.

The contract was signed by Grubb and Fitzgerald, each of whom was identified only as the Seller, and by Clovin, who was identified as the Buyer.

The court concluded that the face of the contract provided substantial evidence that the contract was entered into by Grubb as an individual and that the contract was not between Cloven and any corporate entity. The court found that "The contract refers to the Double Cousins Company only as Agent." Grubb and Fitzgerald signed their names as Seller, respectively, and did not purport to represent anyone other than themselves.[16]

General Conditions

Every construction contract has terms and conditions that apply to all contracts and are not specific to a single project. These general conditions are sometimes included on the same page as the agreement. Some builders include the agreement and the general conditions in one document; others separate the two. The builder should review the general conditions to make sure that they do not conflict with the contract itself.

Obviously the contract cannot cover every contingency. The more time the parties spend identifying their respective obligations before the project begins, however, the fewer surprises will occur later.

Financing and Other Contingencies

If the buyer chooses to use third-party financing, the builder may include a financing commitment from a lender in the contract. The contract should allocate payment of loan application fees and closing costs between the buyer and the builder. The generic term *closing costs*, refers to many fees and expenses. To minimize disputes, the parties should identify and specify who will be responsible for the payment of each cost.

If he or she has not already done so, the buyer should agree to apply for a loan within a certain period after execution of the contract. The contract should specify that if the buyer does not receive a loan, the contract will be terminated and the parties will be released from their obligations under the contract. If the buyer has given the builder a refundable deposit, the contract should specify the conditions under which the builder will return the deposit to the buyer.

A buyer who does not need financing should place funds equal to the total price of the agreement in an escrow account. Similarly, if the buyer needs only partial financing, the balance of the funds due should be placed in escrow.

> The buyer represents that [he or she] has arranged sufficient financing to comply with this agreement. Before commencing work under this agreement or at any time during the progress of the work, the builder may request evidence in writing (acceptable to the builder) of financing for the work.

> Failure of the buyer to produce the requested information within seven (7) calendar days of the written request will constitute a breach of contract by the buyer, and the builder may suspend the work. If the buyer fails to provide the requested information within an additional seven (7) calendar days thereafter, the builder may terminate this agreement. Should the builder suspend or terminate this agreement, he or she shall be entitled to collect or receive payment for materials and labor expended on or for the project, along with a reasonable profit and overhead.

Any other contingencies that the parties contemplate should be specifically identified. For example, if the agreement is contingent upon additional appraisals, soil-bearing tests, percolation tests, inspections, or the sale of other property, the contract should specify the time period in which those events or requirements must occur, what will happen to the agreement if the conditions are not fulfilled, and how the deposit money will be divided.

If the builder sells a lot to the buyer with the understanding that the builder will construct a house on the property, the contract of sale should make the lot sale contingent upon the buyer's execution of a construction contract with the builder within a specified period.

Builders should consult an attorney for specific language for contingencies.

Allowances

The total contract price shall include any allowances mentioned in the contract documents. Contracts frequently provide for allowances for items of work that are not sufficiently detailed in the contract documents to enable the builder to determine the final cost of the item. Their inclusion in the contract establishes that the item is within the builder's scope of work subject to final price adjustment.[17] Allowance items may be of two types: (a) items that remain to be selected by the owner or buyer (such as wallpaper and cabinets) and (b) items of work for which the actual cost cannot be determined until the builder receives additional information regarding the scope of the work, until actual conditions are verified, or until the work is actually performed, such as the drilling of a well. In the first situation the builder should closely monitor the owner's or buyer's spending habits during construction, and periodically notify the owner or buyer of the balance of each allowance. In the latter situation the builder must be able to justify the original estimate because, if the actual cost greatly exceeds the allowance, the builder may have to deal with an angry customer. Moreover, if the allowance was grossly underestimated, a court may allow the builder to recover only the reasonable costs.

The prudent builder will provide in this clause or in the separate allowance schedule that (a) whenever costs are more than or less than allowances, the contract price shall be adjusted accordingly by change order, and (b) the change order shall reflect the difference between actual costs and the allowance. The builder should not assume that the owner or buyer understands that he or she is responsible for paying the difference in cost if the cost of an item exceeds the allowance. The builder shall supply items for which allowances are provided within the amounts specified. Those amounts must cover the applicable taxes as well as the builder's cost for materials and equipment delivered to the site. Trade discounts will be passed on to the owner although discounts for timely payment may be retained by the builder. The allowance item should clearly state whether it is for materials only, installation only, or both materials and installation.

To avoid delaying the work, the owner must promptly select all materials and equipment for which allowances are provided. So long as the builder has no reasonable objection to the vendors selected, the owner may specify the vendors for these items.

Illustrative Case

As work on a project progressed, the general contractor-landlord failed to notify the inexperienced tenants about how much of the allowance had been used. Therefore the court ruled that the general contractor-landlord could not recover all the expenses incurred to "finish out" space occupied by the tenants under a lease even though the tenants exceeded their allowance by $16,270.81. The court held that the landlord was not entitled to recover expenses in excess of the contract allowance from the tenant because in "dealing with neophytes, they had an obligation of fair dealing, which required some notification to defendants of the status of the allowance at a meaningful time."

The court noted that the tenants relied on the landlord to do the job within the allowance. However they did not communicate that expectation explicitly to the landlord. Despite knowledge that tenants often have unrealistic expectations, the landlord did not give anything beyond a subtle warning. The court observed that the case presented the classic perils of a failure to communicate.[18]

Permits, Licenses, and Other Approvals

The builder is usually responsible for obtaining all permits, licenses, fees, and approvals associated with the construction and occupancy of the project. The builder must be careful, however, about agreeing to obtain all permits and the like. He or she is probably better off identifying in the contract the various permits, licenses, fees, and approvals for which he or she will be responsible. For example, the builder who agrees to obtain all permits probably contemplates obtaining the local building permit and the certificate of occupancy permit but does not contemplate obtaining a wetlands permit or a state water quality certification. However if the builder agrees to obtain all permits and the owner's lot contains wetlands, the builder may be required to obtain a wetlands permit before he or she may continue with the project. The permitting process can be lengthy (in some cases more than a year) and expensive.

In addition if the buyer owns the land on which the house is to be built and that lot is in a development that requires approval of plans and specifications, materials, and colors by an architectural review committee, the buyer should obtain those approvals.

> The builder shall obtain and pay for all building and construction permits, licenses, and other approvals necessary for occupancy of permanent structures or changes in existing structures unless the land is owned by the buyer of the house and covenants or a residents' association require approval of plans and specifications, in which case the buyer is responsible for obtaining these approvals and paying for any fees connected with them.
>
> Should the builder fail to obtain the necessary permits, licenses, or other approval for which the builder is responsible, the owner has the right to and shall be entitled to terminate the contract.

Insurance and Risk Management

An agreement to carry insurance coverage is often written into construction contracts. The contract will generally require both parties to buy and maintain insurance for specified injuries or risks and in specified dollar amounts. The contract should clearly state the insurable interest of each party and establish who owns or is responsible for what property, during what period of time, and the type and amount of coverage required.

In new home construction, more often than not, the contract requires the builder to obtain insurance. The need for insurance and the exclusions to insurance policies vary according to the type of construction involved, but some common types of insurance include the following:

- comprehensive general liability
- builder's risk
- workers' compensation
- automobile liability
- professional liability
- completed operations
- umbrella liability

The contract may also require the buyer or owner to obtain insurance. If the builder is building on the owner's land, the builder will want the owner to maintain property insurance that will (a) protect the builder, the subcontractors, and the sub-subcontractors and (b) insure against the loss of work in place or materials on the site. The contract also may require the owner to obtain insurance against loss resulting from injury of third parties (persons not parties to the contract) or to their property.

Builders should consult a construction insurance expert to determine their insurance needs, as well as an attorney for specific contract language.

Risk of Loss

Builders obtain insurance for numerous risks, but one risk that goes to heart of the transaction is the risk of loss caused by accidental destruction of or damage to the house during its construction. The parties should always obtain insurance to cover such a loss. In addition their contract should discuss their respective rights and obligations if such a contingency occurs. Many builders would be surprised to learn that if a builder agrees to construct a complete house, the builder generally bears the risk of loss if the house is accidentally destroyed before its completion, and the builder may be required to rebuild the house. Therefore, if a builder wishes to make some other arrangement in the event the partly completed house is destroyed, he or she would be wise to do so in the contract. For example, a well-known builder provides that if the damage covered by property insurance is greater than 30 percent of the value of the contract, (a) the contract shall be terminated, (b) the builder shall be compensated for work completed before termination, and (c) at the owner's option the contractor will negotiate a new contract for the repair and/or completion of the home. On the other hand, if the loss is less than 30 percent of the contract value, the builder will proceed under the original contract but will prepare a change order for the labor, materials, and profit and overhead costs required to repair the damage. The change order is to be paid from proceeds of insurance and the owner's funds.

Differing Site Conditions

Unexpected site conditions, such as rock or an unexpectedly high water table, can be expensive to work around. This provision protects builders and owners if these site conditions are encountered. It protects builders because, if a builder agrees to construct something under a contract, the builder is not entitled to additional compensation merely because unforeseen difficulties make the job more expensive than anticipated. This clause protects builders because it eliminates the risk of increased cost from unexpected hidden (latent) site conditions. On the other hand, the owner benefits from such a clause because he or she pays for these increased costs of performance only in the event that the builder actually encounters changed or differing site conditions.

The contract should include a clear definition of *differing site conditions* and provide for notification, work changes, or contract modification procedures upon discovery of such a problem. An alternative to the clause provided below could require the buyer to pay the extra costs incurred (plus profit at a specified percentage of the cost). The buyer or builder may also have the option of canceling the contract. If the builder is building on his or her own lot, he or she may reserve the right to switch lots. These suggestions would replace the equitable adjustment language below. The definition of *equitable adjustment* can be confusing. Builders' attorneys should check the case law precedents in their states to see how the term is treated by the courts, and it should be used only where the term has an identified meaning.

> A differing site condition is a physical characteristic of the property that materially changes the construction techniques from those reasonably expected at the time of the contract. Examples of differing site conditions are subsurface or latent physical conditions at the site differing materially from those indicated in this contract or unknown physical conditions of an unusual nature (that are not reasonably foreseeable) on the building site.
>
> Before disturbing any differing site condition, the builder shall notify the buyer of such a condition except in the case of eminent danger to persons or property. The buyer shall investigate the condition within five (5) business days. If the buyer and the builder agree that the differing site conditions will cause an increase in (a) the builder's cost of performance of any part of the work under this contract or (b) the time required for that work, the issue will be resolved as follows:
>
> A. If the total contract price will increase by more than ________ percent (____%) [for example, 10 percent], the buyer may terminate the contract upon paying the builder for all costs expended to date and for the builder's full profits as provided in Section ____, Price, Deposit, and Payments, of this contract; or
>
> B. The builder and the buyer shall execute a written specific cost adjustment to this contract, including any adjustments in the time for performance required by the differing site conditions.
>
> The buyer's failure to investigate the condition will confer authority upon the builder to complete construction of the project according to the terms in Section ____, Liquidated Damages and Unavoidable Delay, of this contract, and the builder shall further be entitled to [an equitable adjustment or payment of any increased costs necessitated by the differing site condition].
>
> If the parties cannot agree on the existence or consequences of a differing site condition, the terms of this provision shall be arbitrated as provided for in Section ____, Arbitration, of this contract.

Architects and Construction Managers

Sometimes buyers hire architects or construction managers who participate in the construction of their houses. The contract should name the architect or construction manager and designate whether or not he or she is the buyer's agent.

Warning—A builder should not rely on an architect or on a construction manager unless he or she has written authorization to act on behalf of the owner. If the contract provides such authorization, it also should specifically provide that the builder will not be held liable for any actions made in reliance on recommendations of the architect or construction manager.

The architect or construction manager typically is responsible for—

- inspection or observation
- processing change orders
- overseeing the payment process
- interpreting or obtaining interpretations of plans and specifications pertaining to design considerations.

Of course, the architect or construction manager cannot change the obligations of the parties as spelled out in the contract. Sometimes the architect or construction manager has a role in disputes. The parties could agree that before they can submit a dispute for arbitration or litigation, for instance, they must submit it to the architect for consideration as a mediator.

If applicable, the contract should name the buyer's real estate agent.

> This contract specifies the terms between (buyer's name), the buyer, and (builder's name), the builder, to [sell and/or construct] a home on the property located at: (legal description of property).
>
> [Name of architect and/or construction manager] is the buyer's agent, and the builder may rely on representations, statements, revisions, and approvals made by [architect and/or construction manager] related to the performance of the terms of this contract. The builder will not be held liable for any actions made in reliance on recommendations of [architect and/or construction manager].

Changes in Work and Change Orders

The buyer does not have the automatic right to order changes in the work unless the contract confers that right. However most contracts include a change order clause giving the buyer that right. This section sets up the procedure for writing change orders and explains how the cost will be determined, including overhead and profit. The actual change order is a separate document (Figure 2-3).

After the contract is signed and work commences, often the buyer will request that changes be made in the design or materials used. Before beginning any new work created by a buyer's request for a change, the builder should require that the buyer execute a written change order and pay for the work. A change order is an agreement that should specify (a) the revisions in the work and the price, (b) a revised payment schedule, if necessary, and (c) a new date for substantial completion, if necessary.

If the parties cannot agree on the price of the change order work, the contract could include a provision for the builder to be paid for work done on a cost-plus basis. If the contract includes more than one owner or purchaser, getting both owners to sign the change order may be time-consuming. To expedite the work, the contract might provide (a) that either owner may sign the change order as agent for the other and (b) that the signature of one owner is binding on the other owner.

A change order should address all changes that affect the scope of the work, the contract price, and the time for performance, not just changes in the work requested by the owner. Thus additional work to be performed under a differing site or concealed conditions clause should be handled by a change order. Similarly any increase in the scope of the work required by the building and/or planning department that is not the builder's fault should be handled by a change order.

The contract should expressly provide that such an increase will be treated as extra work and that the extra work will be done after a written change order is signed by the buyer. Generally the contract will call for payment of each change order to be made (a) when the work is performed, (b) upon completion of the change order, or (c) before the next draw.

A builder who performs work without obtaining written change orders and who then presents the buyer with a large bill for that work at closing is asking for trouble.

Figure 2-3. Sample Change Order

Change order number ________________

Date ______________________________

Project description __________________________

Project number__________________________

Description of change (including reference to drawings and specifications revised, new drawings and specifications issued)____

____

____

Reason for change

____

____

____

Total price prior to this change $ ______________

Change in price for this change order $ ______________

Total revised price $ ______________

Revised schedule of payments:

____

____

____

____

The estimated completion date provided for in paragraph ____ of the contract is now __(date)__. All other terms and conditions of the contract referred to above remain unchanged.

______________________________ (builder)

______________________________ (buyer)[1]

Date ______________________________

Date ______________________________

1. This form is designed for a single buyer. If more than one buyer is involved, the form should be adapted to accommodate the initials and signature of each of the buyers.

The buyer may request changes in the work within a reasonable scope. Upon written directive by the buyer, the builder will make changes, additions, or alterations. If the buyer and the builder agree on the cost of the modifications, they shall sign a written change order describing the changes to be made, any extra work to be done, and any changes to the contract price or completion date (Figure 2-3). Change orders shall be signed by all parties and become part of this contract, and the buyer agrees to pay the builder for changes in the work on the same basis as specified in Section ____, Price, Deposit, and Payment Provisions, of this contract. [If more than one buyer is involved add the following: The buyers agree that either of them may sign the change order and that the signature of one is binding on the other.]

If the buyer and the builder cannot agree on a fixed cost of the change, the builder will make the changes and will receive payment from the buyer. Payment will include the costs of labor, materials, equipment, and supervision plus ______________ percent (_____%) of such costs. The buyer agrees to make requests concerning any changes, additions, or alterations in the work in writing directly to the builder named in this contract and not to the workers, including subcontractors and subcontractors' workers on the job.

Illustrative Case

A builder of low-rent housing who submitted change orders and performed extra work authorized orally by the developer's architect could not recover for the cost of extra work under its contract with the property owner because the change orders were not approved by the owner as specifically required by the contract. The developer who accepted the work was not a party to the contract and did not approve the changes in writing. Fortunately for the builder, the court allowed recovery on other grounds.[19]

Mechanic's Lien

Mechanic's lien laws (construction lien laws in some states) ensure that participants in the construction process get paid for their work by granting them a specific interest in real property that has been improved by their labor or materials. All 50 states, the District of Columbia, and Puerto Rico have such laws. The requirements of these laws vary considerably from state to state. For a builder, subcontractor, or supplier to benefit from or comply with a lien law, the procedures and requirements of the law must be strictly followed.

The claimant (the person claiming the lien) must provide one or more forms of notice before the lien can be effective. This process is often referred to as *perfecting the lien* (language that is commonly used in the mechanic's lien statute). To perfect the lien often the claimant must file a formal *Claim of Lien*. Such a filing must be made within a fixed period after completion of the contract or the last furnishing of services or materials. When the lien has been perfected, it may be enforced in a lawsuit to compel the sale of the property. The time for enforcing the mechanic's lien varies from state to state, but seldom exceeds 1 year after the date on which the lien was filed.

A majority of states allow a builder to waive his or her lien rights by signing a lien waiver before beginning work for the owner or buyer. The lien waiver clause typically provides that the builder will not file any liens against the property on account of labor or material or equipment furnished pursuant to the contract. A lien waiver clause offers little, if any, benefit to a builder. Homeowners find such a provision attractive because it ensures that the property will remain unencumbered by liens.

Obviously builders should be wary of signing any form which has the effect of waiving their lien rights before they receive payment, and they should be sure they can recognize a lien waiver and be on the lookout for such wording because the owner's attorney or the lender may suggest adding it.

> The builder unconditionally waives, releases, and relinquishes all right to file or maintain any mechanic's lien or other claim in the nature of a lien against the real property improved under this contract or the building on account of any labor, material, equipment, extras, change order work, or increased costs.

Inspection, Acceptance, and Possession

Inspections allow buyers to identify, and give builders the opportunity to correct, defects in materials or workmanship. Because the buyer or the buyer's agent makes periodic inspections, the builder can document a buyer's satisfaction with the completed home. The builder should keep a record of each inspection for (at minimum) the duration of any statute of limitations or repose (see Figure 7-1, Sample Home Maintenance Instruction Checklist, and Figure 7-2, Sample Punchlist).

Statutes of repose and statutes of limitation limit the time within which an injured party may sue the person who caused the injury. The statutes differ with regard to when the "clock begins to tick." For a statute of repose it begins to run at an arbitrary point in time, such as when the building is substantially completed, and usually before the person suffers a loss. Conversely the time allowed by the statute of limitations begins to run when the injured party discovers or reasonably could have discovered the problem. In practical terms from the builder's perspective, the difference between the two is that without the statute of repose the builder could be liable for the entire life of the structure.

The contract should define *possession* and determine what steps the parties must make before the buyer takes possession. For example, builders often include provisions for inspection, formal acceptance, and final payment before a buyer can take possession. Some contracts state that (a) the owner or buyer will not occupy the house until final payment of all sums due, including any extra charges, and (b) occupancy of the home shall be deemed acceptance of the work, including defects that could have been discovered upon a reasonable inspection by the owner or buyer.

The sales contract should state that "time is of the essence," and require the builder and buyer to make full settlement of the terms of the contract by a specific date (Figure 2-4). This certificate must be part of the contract documents because the buyer must see this document before the job is finished and have an opportunity to ask questions about it. If the buyer sees it for the first time when asked to sign it, he or she is likely to balk at doing that.

> From time to time and just before the closing, the buyer or the buyer's agent shall inspect the house in the presence of the builder or the builder's representative. At final inspection, the buyer will give the builder a signed punchlist that identifies any alleged deficiencies in the quality of the work or materials.
>
> The builder shall correct any items on the buyer's punchlist that are, in the good faith judgment of the builder, deficient in the quality of work and/or materials according to the standards of construction relevant to the jurisdiction in which the house is built. The builder shall correct those defects at its cost within a reasonable period of time. The builder's obligation to correct any defects shall not be grounds for postponing or delaying the closing, nor for imposing any conditions upon the closing not specified in this contract.
>
> Upon substantial completion of construction, the buyer shall sign the certificate of acceptance (attached and incorporated by reference in this contract). The buyer shall not take possession of the project before inspecting the home, signing the certificate of acceptance, and making the final payment to the builder. This provision is subject to the right of either party to obtain an injunction or file for any other legal or equitable remedies.

Representations and Warranty

The builder should discuss the limited warranty with the buyer at the beginning of the negotiations (see Chapter 6). Most courts will uphold limitations on or exclusions to the warranty if they are considered part of the contract. That is, when warranty limitations and exclusions are included in the give and take between the builder and the buyer when they are forming their agreement, the courts will usually enforce these limitations and exclusions.

But many courts also require that to be enforceable the disclaimer of implied or other warranties be conspicuously displayed in the contract. The disclaimer should appear either on the first page of the contract or just above the buyer's signature. Putting the disclaimer in bold type also may help. Finally builders should require the buyers to initial this section separately.

As discussed in detail in Chapter 6, in a majority of states, professional builders impliedly warrant that the houses they build and sell are (a) habitable and (b) conform to the building standards then prevailing in their community. As noted above in some states these warranties can be disclaimed if certain notice requirements are satisfied.

Builders can assign the manufacturers' original purchase warranties directly to their buyers. Assigning warranties transfers directly to the buyers those warranties that the builders received when they purchased the goods for installation into the buyers' new homes.

Figure 2-4. Sample Certificate of Acceptance

The buyer certifies that all of the terms and conditions of the contract entered into by the buyer and the builder for the [construction and/or purchase] of property at (address of house) have been met, and the buyer and the builder further acknowledge and agree as follows:[1]

1. While the house has generally been constructed according to basic plans and specifications contained in the contract documents, the buyer understands that the house may not correspond in some respects with those plans and specifications because changes may have been made before or during construction. These changes may be attributed to a variety of events, including changes in topography, construction techniques, building codes, the availability of material, or other events.

The buyer understands that the builder is not obligated to furnish any as-built plans, specifications, or drawings of the house. The buyer also understands that the house may differ in some respects from the models, drawings, maps, pictures, or other depictions of the house the buyer was shown. The buyer acknowledges that minor variations may exist in the outside and inside dimensions, configurations, colors, location, general appearance, and other characteristics. The buyer has had an opportunity to inspect all aspects of the project and has done so or elected not to do so.

2. The buyer has inspected the house and the property on which it is located. The buyer has also delivered to the builder a written list of all items the buyer believes (a) have not been properly constructed or (b) are not in proper condition and has described the specific problems. Except as noted on the list, the buyer accepts the residence and property as is, and acknowledges that from now on the buyer will have no claim against the builder for any item that was not listed that could reasonably have been ascertained or observed during the buyer's inspection. The buyer has no objections relating to color, appearance, type or brand of equipment, dimension or size, location, breakage or cracks, or any other conditions that reasonably could have been discovered by the buyer during the inspection.

3. The buyer understands that no warranties are being made by the builder except those appearing in the written limited warranty provided by the builder as part of the contract documents. All statements, representations, promises, and warranties made by the builder or any agent of the builder are superseded by the written limited warranty, and the buyer is not relying on any representations, promises, or warranties except for the written limited warranty that is included by reference in this acceptance document.

The buyer understands that the duration of all implied warranties from the builder, including (but not limited to) the implied warranties of habitability and workmanlike construction, have been limited by the builder to one (1) year from the date of sale or the date of occupancy, whichever comes first.[2]

4. The buyer understands that in exchange for the limited warranty and the other provisions of the contract with the builder, the buyer will have no right to recover or receive compensation for any incidental, consequential, secondary, punitive, or special damages, nor any damages for aggravation, pain and suffering, mental anguish, or emotional distress, nor any costs or attorney's fees. This provision shall be enforceable to the extent allowed by law.

5. The buyer and the builder agree that the claim procedures described in the limited warranty and the arbitration procedures described in the contract shall apply to any claims made by the buyer or the builder, and the binding arbitration provisions of the contract shall be the sole available remedy for any unresolved dispute relating to the construction of the house.

6. The buyer acknowledges satisfaction with the manner in which the sales transaction was closed, including all financial calculations and adjustments. And buyer acknowledges receipt of all closing documents to which the buyer is entitled from the builder.

7. Each provision of this certificate is separate and severable from every other provision. If any single provision is declared invalid or unenforceable, the buyer and the builder understand that all the other provisions will still be valid and enforceable. The buyer and the builder agree that every provision of this certificate will survive the closing of the sales transaction and will not be merged with the deed.

______________________ (buyer's signature)

______________________ (name of builder, corporate name where applicable)

By ______________________ (authorized signature)

Date ______________________

Title ______________________

Date ______________________

1. Sample language that appears in brackets presents an option for the reader. Depending on the particular transaction the reader must choose one or the other, both, or neither of the possible wordings presented. A builder may wish to simplify this certificate of acceptance by eliminating or reducing the language in paragraphs 1, 3, 4, and 5, which largely repeat provisions already covered by sample language in this chapter. This form is designed for a single buyer. If more than one buyer is involved, the form should be adapted to accommodate the initials and signature of each of the buyers.

2. Builders should make sure that this limitation of the implied warranties conforms to any limitation appearing in the limited warranty (see Limited Warranty in Chapter 6 and Representations and Warranty in this chapter).

Warning—Some manufacturers do not allow assignment of warranties on their products, and the assignment of warranties is governed by state law.

If builders intend to warrant the manufactured items or consumer products installed in the homes they build, the warranty must conform to the Magnuson-Moss Act guidelines (see Chapter 6 for more information on this federal law).

The warranty limitations should also address environmental issues, such as indoor air quality, including radon, and possibly fiberglass (see Chapter 5 for a full discussion of these and other environmental liability issues).

Builders must instruct all agents, representatives, and employees not to make any promises or representations to buyers without specific authorization from the builders. In addition the contract should provide that the builder has made no guarantees, warranties, understandings, or representations (and that none have been made by any representatives of the builder) that are not set forth in the contract and the warranty. This section should be signed or initialed by the buyer or owner.

The builder's jobsite supervisors, managers, employees, and subcontractors who are likely to be in contact with the buyers should sign agreements with their individual builders acknowledging that they are not authorized to make representations, promises, warranties, and the like without authorization of the builders.

Access

In the interest of safety and to minimize the buyer's or owner's opportunity to communicate with subcontractors regarding the means, method, or manner in which they are to perform their work, some contracts limit the time of day during which the buyer or owner may visit the site. Requiring the owner or buyer to schedule visits with the builder may prevent injuries that might otherwise occur if the owner or the buyer is free to roam the site unaccompanied. For example, the contract might provide for site visits in the presence of the builder early in the morning or in the evening. Alternatively the builder may give the owner or buyer free access to the property with the agreement that the owner or buyer enters the property during construction at his or her own risk.

In either case the contract might provide that the owner or buyer must indemnify the builder against claims, damages, property damage, or bodily injury arising out of the owner or buyer's unaccompanied site visit, but only to the extent it was caused in whole or in part by the owner or buyer's negligent acts or omissions. Obviously, because the buyer or owner is paying for the home, the builder should be sensitive to (and respectful of) the buyer's or owner's right to participate in the building process by visiting the site, asking questions, and commenting on the work. The builder may have more control over the owner's or buyer's access to the site if the builder owns the lot because keeping the owner off his or her own property may be difficult.

> The owners shall at all times have access to the property and the right to inspect the work. However, if the owners enter the property during the course of construction, they do so at their own risk, and the owners hereby release the builder and do hereby indemnify and hold the builder harmless from any and all claims for injury or damage to their person or property, and to the person or property of any agent, employee, or invitee of the owners or of any other person accompanying the owners.
>
> The owners shall not in any manner interfere with work on the job nor with any subcontractor or workers. The owners will not communicate directly with the builder's workers, employees, agents, or subcontractors regarding the means, method, or manner in which they are to perform their work. If the owners delay the progress of the work, causing loss to the builder, the builder shall be entitled to reimbursement from the owners for such loss.

Work Performed by Owner and Other Contractors

Some buyers and owners reserve the right to perform some of the work themselves or they reserve the right to subcontract part of the work. Some builders strongly believe that the owner should not perform any of the work, and they try to discourage this practice. They may even include such a warning in the contract.

If the owner reserves the right to perform some of the work or to award separate contracts in connection with other portions of the work, the contract must clearly describe the work to be performed by the owner or the subcontractor, including the time within which the work is to be performed.

If the owner enters into multiple contracts with different contractors for work on one project, the owner generally is under an obligation to coordinate and control the operations of all contractors to avoid unreasonable disruption of, or interference with, the operations of the other contractors. If that coordination fails the delayed performance may be compensable. Finally, if the owner performs certain work or has other contractors perform that work, the builder should expressly provide that the builder is not warranting such work.

Conveyance

This provision applies only to the sale of property. When a builder transfers property and, by necessity, the deed to the property, the builder will make certain warranties about the condition of the title to the property. The most common warranties are included in a general warranty deed. This type of deed assures the buyer that the seller has transferred specific exemptions for commonly recorded information about the property such as covenants, easements of record, municipal and zoning ordinances, and any other liens or encumbrances specified in the agreement.

Alternatively builders can supply special warranty deeds, in which the sellers warrant that they did nothing to cloud the title. Or they can provide a *quit-claim deed*, which provides no warranties to the buyer. A quit-claim deed simply releases any ownership rights that the seller may have in the property. Builders should consult an attorney to determine the prevailing practice in their communities and for specific language.

Mandatory Clauses

Both builders and remodelers must take care to include certain mandatory language in their contracts. Several federal laws require that certain notices, warnings, or other information be included in new home sales contracts or consumer contracts (which often include remodeling contracts). Similarly state laws often require that certain information be included in custom home contracts, new home sales contracts, or consumer contracts. For example, the Federal Trade Commission requires all builders and sellers of new homes to disclose to customers information about the type, thickness, and R-value of the insulation installed in the house. And under Maryland law a custom home builder must include in each custom home contract a disclosure concerning the buyer's risk under mechanic's lien laws.[20]

> **Insulation**—Pursuant to federal law, a new home sales contract must include the following information: the type, thickness, and R-value of the insulation to be installed in each part of the house. This information may be provided at a later time, however, if the buyer signs the sales contract before the seller knows what type of insulation will be put in the house.[21]

Escalation Clause

If a party agrees to perform work for a fixed price, that party bears the risk of an increase in the cost of compliance because of an increase in the cost of labor or materials during the project. In other words a party will not be excused from completing the work covered by the contract just because it turns out to be more difficult or burdensome to perform. The ongoing saga of the increase in lumber prices is evidence that an increase in the cost of materials during the project can significantly increase the cost of compliance of a fixed-price contract.

The parties can provide in the contract that if a contract price increases because of any price increase in labor or materials or because of additional work required by the building department or other governmental agency, before or during construction, the owner will pay the builder the increase upon proof of such an increase. In addition the parties might agree that if a price increases more than a particular percentage, the owner has the option of paying the increase or terminating the contract after reimbursing the builder for all work completed before notice of termination.

Illustrative Case

In this case, the Hudsons sued D and V Mason Contractors, Inc., for breach of contract to construct a house when the builder refused to complete the Hudsons' house for the agreed-upon price of $17,400.00. The building company contended that it was excused from performing because of a scarcity of labor in the building industry and an increase in

(a) the interest on construction financing, (b) the points the builder was required to pay on the mortgage, and (c) construction costs.

The builder relied on a provision in the contract that made the builder's ability to complete the house dependent upon conditions similar to those existing at the time the contract was signed. The contract gave the builder the option to cancel the contract if the builder was unable to promptly obtain the labor and materials required for construction as needed. The builder could also cancel the contract if any present or future rules, regulations, or restrictions by the federal, state, or municipal governments prevented the builder from completing the project.

Finding for the owners, the court cited the general rule that a court cannot alter a contract merely because it will create a hardship. The court noted that contractual provisions attempting to qualify an undertaking in the face of increased cost considerations must be specific with reference to those factors that will excuse performance. The court concluded that the three cost factors cited by the builder were not grounds to excuse performance within the meaning of the contract and that the builder failed to show that it was particularly and substantially affected by the scarcity of labor.[22]

Clean-Up

Industry practice requires the builder to leave the completed job in a clean condition. The inclusion of cleaning in the contract makes the builder's obligation clear.

> Upon completion of the project, all of the builder's construction debris and equipment shall be removed by the builder, and the premises shall be left in neat, broom-clean condition, unless otherwise agreed upon herein.

Signage

Projects that are run in an orderly manner represent one of the best advertisements a builder can have. A sign displayed at the job can produce more work for the builder. Although common practice allows a builder to display a sign at the project, contract language to that effect will make that right clear.

> The buyer agrees to permit the builder to display a sign on the site until the project is completed.

Uncovering and Correcting Work

Other topics typically included in general conditions include uncovering and correcting work, for example, work that (a) was covered before inspections occurred, (b) does not meet the contract standards, or (c) violates some other aspect of the contract. The general conditions can also provide for acceptance of work that was not done to contract specifications but that is still acceptable to the client. The following sample language could also be adapted by the builder for use in the subcontractors' contracts to protect the builder.

> **Uncovering Work**—If work is covered in contradiction to the contract or the applicable laws, the builder shall uncover the work for observation or inspection and replace it without charge. If the home buyer or the architect request that work be uncovered, the builder shall obtain a change order signed by the home buyer to charge the home buyer for uncovering work that has been completed and inspected in compliance with the contract and local laws. If the work uncovered proves not to comply with the contract documents or applicable laws, the builder shall bear the costs of the uncovering and redoing of the work unless the home buyer caused the problem that needs to be corrected.
>
> **Correcting Work**—Upon written notice by the buyer, the builder shall promptly redo and recover work that does not meet contract specifications or applicable laws and pay the costs of redoing that work and recovering it as well as any additional expenses for testing or inspections.
>
> If within a reasonable time the builder does not correct work that does not meet contract requirements or applicable laws, the buyer or owner may remove it and the builder's equipment, and store the equipment and materials that can be reused, at the builder's expense. If the builder does not pay the costs of removing and storing the equipment and materials within a specified time, the buyer can sell the equipment and materials. If the proceeds of the sale do not cover what the builder would have paid to correct the work, the contract price shall be decreased accordingly. If the noncomplying work destroys or damages work done by the owner or another contractor, the builder shall bear the cost of correcting that work as well.
>
> **Acceptance of Nonconforming Work**—Instead of requiring removal of the work, the buyer may choose to accept work that does not meet contract specifications so long as it conforms to the

applicable laws. If the owner chooses to do so, the total contract price will decrease accordingly regardless of whether the buyer has made the final payment.

Termination of the Contract

Occasionally unpredictable or uncontrollable events substantially change a construction project, and one party wants to be released from the contract. The builder and the buyer can negotiate how to allocate responsibility for this possibility. Some of the terms in this clause will depend on the builder's business philosophy. For example, a well-known custom builder allows for termination at the convenience of the owner, so long as the owner pays him for all labor and materials furnished plus a proportional share of the overhead and profit associated with the job. This builder is confident that such an event is unlikely to occur, but he is also comfortable giving the owner the power to terminate the contract.

Such a termination clause is unusual. Most contracts provide for termination for cause, such as the owner's nonpayment, the builder's substantial failure to comply with the contract terms, a substantial delay caused by a government entity that interrupts the work, or a substantial cost increase resulting from the discovery of a hidden or unforeseen condition. In the absence of such a clause and unless termination is the result of mutual agreement, a party may not terminate the contract without liability unless the other party defaults and the default goes to the root of the contract—often referred to as a *material breach of the contract*.

The termination clause should give the parties the right to cure their default. For example, the contract may include a provision that allows the buyer to cure a breach of contract within a specific time by bringing all payments and other obligations up to date. This action permits the parties to reinstate the agreement so the builder can complete the job. Similarly if the contract authorizes the owner to remove the builder from the job because of deficiencies in the work, the builder will include wording in the contract stating that the builder is entitled to notice of the deficiency and an opportunity to correct the deficiency before the termination is effective.

The contract should require that all termination notices be in writing (see Notices).

Builder's Remedies—If the buyer materially fails to comply with the provisions of this contract or terminates the contract for any reason other than the builder's failure to perform, the builder may (a) terminate this contract and retain any downpayments and/or deposits as liquidated damages; (b) recover all unpaid costs, expenses, and fees earned up to the time of default or termination; the prorated cost of overhead expenses; and the costs, fees, and prorated overhead expenses for all change orders approved by the buyer before the termination; and (c) institute judicial proceedings for specific performance and/or any other legal and equitable remedies. (The costs specified in this section shall be based upon the costs specified in Section ____, Price, Deposit, and Payment, of this agreement.)

Buyer's Remedies—If the builder fails to supply proper materials and skilled workers; make payments for materials, labor, and subcontractors in accordance with their respective agreements; disregards ordinances, regulations, or orders of a public authority; or materially comply with the provisions of the contract, the buyer may give the builder written notice. After seven (7) days if the builder has failed to remedy the breach of contract, the buyer can give a second written notice. If the builder still fails to cure the breach within seven (7) days after the second notice, the buyer may terminate the contract. When the buyer terminates the contract for one of the reasons stated above, he or she may be entitled to seek legal and equitable remedies.

Arbitration, Mediation, and Other Alternative Dispute Resolution

During the course of construction, a disagreement might arise between the builder and the home buyer or lot owner that they cannot resolve through negotiations. Because of the expense of a lawsuit and the long wait to have a case heard in court, litigation may not be a wise or efficient method of resolving the matter. The contract should therefore address how disputes will be settled. Builders can use several methods of resolving disputes short of litigation. Alternative dispute resolution (ADR) methods include the popular arbitration and mediation techniques.

For flexibility, builders should consider including a clause in their contracts that allows them to choose arbitration, mediation, or litigation.

Arbitration

Binding arbitration is a process in which the parties submit their case to a neutral third person or a panel of individuals (arbitrators) for a final and binding resolution. Arbitration provides a mechanism for resolving disputes without the publicity of a lawsuit and usually at a lower cost. Arbitrators generally are experts and therefore may be more likely than a jury to understand the technical aspects of a construction controversy. Arbitration also often provides a speedier resolution than litigation. Additionally the mere existence of an arbitration provision may deter potential lawsuits.

States vary in their willingness to enforce arbitration provisions, but the clear trend is to uphold an arbitration agreement. If materials for a house cross state lines, the Federal Arbitration Act (FAA) may govern. Both the FAA and the individual state acts provide that if the contract has an arbitration clause, one party may force the other to arbitrate. Builders should suggest arbitration or mediation procedures early in a dispute, before a buyer begins to suffer mental anguish that a jury might find worthy of compensation with a damage award.

In a minority of jurisdictions an arbitrated case can still be brought to court, even if a contractual provision states that the arbitrator's decision is binding. In the majority of states, however, the arbitrators' decision is final and binding, and neither party can appeal it except in the case of fraud on the part of the arbitrator. Although the parties in these jurisdictions may attempt arbitration before filing a lawsuit, if one party decides to circumvent the process, a court may allow it.

A formal dispute resolution organization such as the American Arbitration Association (AAA) can conduct an arbitration proceeding. The AAA follows its *Construction Industry Arbitration Rules.*

An AAA proceeding under these *Rules* can be expensive because the parties pay both the AAA and the arbitrator(s) of the dispute. Generally the AAA assigns three arbitrators for disputes over $100,000 and only one for cases under $100,000.

Alternatively the parties can provide tailor-made arbitration rules either in the contract or in a separate contract document and thus eliminate the need for an arbitration association. For example, the parties can agree that each party to the contract independently would select an arbitrator, and those arbitrators jointly would choose a third.

The contract also may—

- Provide for discovery and application of the rules of evidence. During formal discovery each side unearths facts and documents from the other side that may be helpful in defending or prosecuting its case. Discovery includes depositions, interrogatories (written questions and answers), and production of documents. Usually discovery is too expensive and time-consuming.
- Set time limits for presentation of each party's case.
- Limit the maximum damages allowed in the arbitrator's decision and award.

Rather than develop their own dispute resolution procedures the parties can agree to follow the procedures outlined in the federal Magnuson-Moss Warranty Act, 15 United States Code, sec. 2301 et seq. (see also 16 Code of Federal Regulations sec. 107).

Builders should not mention alternative dispute resolution in their warranty provisions because they may be forced to use the Magnuson-Moss settlement procedures.

The AAA can be contacted at the address listed below:

American Arbitration Association
140 West 51st Street
New York, NY 10020
(212) 484-4000

(See related discussion of statutes of repose and statutes of limitations in the preceding section on Inspection, Acceptance, and Possession.)

All disputes between the parties to this contract arising out of or related to any contract term(s) or any breach or alleged breach of this contract will be decided by arbitration unless the parties mutually agree otherwise in writing.

The arbitration shall be conducted by (specify the organization or named arbitrator(s) agreed upon) in accordance with [the rules adopted by the arbitration body chosen, the rules specified below, or the alternative dispute resolution proceeding specified in the federal Magnuson-Moss Warranty Act, 15 U.S.C.A., sec. 2301 et seq. (see also 16 C.F.R. sec. 107). The parties reference the Magnuson-Moss

Act only to provide rules of arbitration in the event of a dispute, and they specifically do not incorporate the Magnuson-Moss warranties into this contract.]

The parties must file a written notice of arbitration with the other party to this contract and with [the arbitration association or arbitrator(s) chosen]. The notice of arbitration may not be filed after the date that a claim based on the dispute would have been barred in a judicial proceeding by the applicable statute of limitations or repose (cessation of activity).

Either party may specifically enforce (a) a decision rendered under this agreement to arbitrate or (b) any valid agreement to arbitrate with additional persons, under applicable arbitration laws. The award rendered by the arbitrator(s) will be final and binding, and any court with jurisdiction over the decision may enter a judgment upon the arbitrator's decision.

Mediation

Like arbitration, mediation is a process whereby the conflicting parties meet voluntarily to negotiate a private and mutually satisfactory agreement aided by a neutral third party. A key difference between the two methods is that unlike the arbitrator, the mediator does not make a decision in favor of one party or the other. Instead a mediator focuses on negotiation and problem solving. The mediator assists the parties in this process. Builders' mediation may follow the *Construction Industry Mediation Rules* of the American Arbitration Association.

If mediation has a down side, it may be that, if mediation is not successful, the parties are back where they started and may have to resort to litigation or another form of alternative dispute resolution. For this reason the parties may want to include a provision in the contract making all communications during the mediation confidential.

If a dispute arises between the parties relating to or arising out of any provision of this agreement or any breach or alleged breach of this contract, either party may request mediation. Either party may invoke the dispute resolution procedure of this clause by giving written notice to the other. The notice should include a brief description of the disagreement. A mediator will be selected by mutual consent of the parties. The parties agree to participate in good faith in the mediation to its conclusion as determined by the mediator. No party will be obligated to continue in the mediation if a resolution has not been reached and put in writing within (number of days) of the first mediation session. The costs of mediation, including fees and expenses, shall be borne equally by the parties.

At this point the clause would include the sample language from the arbitration clause printed above beginning with the phrase, "No arbitration proceeding under this provision . . ." and ending with the phrase "and any court with jurisdiction over the decision may enter a judgment upon the arbitrator's decision."

Attorney's Fees

The general rule in this country is that each party to a lawsuit pays its own attorney's fees—win or lose. Thus the prevailing party in a lawsuit often ends up paying a considerable amount of the judgment to his or her attorney or pays a considerable sum of his or her own money to the attorney for successfully defending a suit. One way the parties can handle such an expense is to contract for the award of attorney's fees to the prevailing party in any action arising out of the project. Where the contract includes such a clause, the decision to pursue a case may be made strictly on the merits of the case, whereas if the contract is silent on this matter, the party must factor in the cost of the attorney's fees in deciding whether to pursue the claim. Similarly the inclusion of such a clause may discourage frivolous suits and may force the parties to deal more forthrightly with each other. On the other hand some attorneys do not favor including such a clause in the contract because they fear that it will have the opposite effect and will encourage litigation.

If either party needs to enforce provisions of this contract or to obtain redress for the violation of any provision hereof, whether by litigation, arbitration, or otherwise, the prevailing party shall be entitled to any reasonable attorney's fees, court costs, or other legal fees incurred herein in addition to any other recovery obtained in such action.

Entire Agreement and Severability

This provision emphasizes that this contract supersedes all previous agreements between the parties. It also assures that if a court determines that one provision is unenforceable, the remainder of the contract will remain in effect. The merger clause, necessary only where the builder is selling the land or lot on which the house is built, specifies that the provisions of the contract will not become void upon transfer of the deed.

Finally the parties can agree not to assign or sell their rights and responsibilities in the contract to another party. The assignment clause restricts the buyer's right to sell the contract (and its payment obligations) to another person who may not have adequate financial resources to pay the builder.

> This contract (including the documents incorporated by reference herein) constitutes the entire agreement between the parties. It supersedes all previous or contemporaneous agreements and understandings, whether written or oral. This contract may be changed only by a written document signed by all parties to this contract. The contract shall be binding upon each of the parties' respective heirs, executors, administrators, successors, and assigns. This contract shall not be assigned, however, without the written consent of all parties.
>
> Each provision of this contract is separable from every other provision of the contract, and if any provision is unenforceable or revised the remainder of the contract will remain valid and enforceable.
>
> This contract will be governed by the law of the jurisdiction in which the property is located.
>
> The provisions of this contract shall survive the execution and delivery of the deed and shall not be merged therein.

Specifications

The general building specifications should be prepared as a separate document and referenced in the agreement along with other documents such as plans and drawings. The specifications are contained in a written document that describes the work to be performed (construction specifications) and the materials to be used and how they are to be installed (material or product specifications). The parties to the contract should initial every page of the specifications and plans, in order to preclude a charge that a different page has been substituted or a plea of ignorance in the event of a disagreement between the builder and the owner.

The specifications should be as detailed as possible. They should reference plans and drawings where appropriate and include provisions for the items listed in Figure 2-5. This list is provided only as an example. Other necessary parts of the construction process may not be listed here, and some of the listed items may not apply to a particular project. All work should be listed and described on the specifications document. Wherever possible the material specifications should include size, color, material, quantity, construction, style, brand name, model number, and other descriptive information.

The construction specifications should be equally detailed. For example, the specifications for painting should read something like, "Painting, Exterior—One coat prime __[brand name, color, and latex or oil-based, if the parties have agreed on these items]__; __[one or two]__ coats finish __[brand name, color, and latex or oil-based, if the parties have agreed on these items]__."

The commonly accepted format for residential specifications is the four-page Description of Materials used for application for mortgage insurance by the Federal Housing Administration, the Department of Veterans Affairs, and the Farmer's Home Administration. It appears in Figure 3-12, Sample Specifications/ Takeoff in *Estimating for Home Builders*.[23]

Description of Work to be Performed

The cost of the work and the builder's fee are based on the work described here. To avoid disputes about whether certain work is required by the contract or is outside of and entirely independent of the contract, this section and the specifications must clearly describe the work to be performed under the contract. The distinction is important because unless the contract otherwise provides, the owner has no right to order the contractor to perform work outside the scope as defined by the contract documents, and the contractor will not normally perform it except in exchange for extra compensation.[24] This section specifically defines the scope of the work in the following ways:

- specifies what the builder will do and, with respect to certain items, what the builder will not do—particularly items that the builder may have reason to believe the owner or buyer is expecting

Figure 2-5. Items Included in Specifications

- Excavation and land conditions
- Footings and foundation or slab (including moisture barrier and termite treatment)
- Chimneys
- Fireplaces
- Exterior walls
- Floor framing
- Subflooring
- Partition framing
- Ceiling framing
- Steel columns and girders
- Roof framing and sheathing (including built-up roofing and roofing materials)
- Gutters, downspouts, flashing, and other sheet metal work
- Masonry or frame exterior walls
- Insulation
- Drywall or lath and plaster
- Windows, glazing, and mirrors
- Interior doors and trim
- Entrances and exterior detail (including entrance doors and trim)
- Cabinets, countertops, backsplashes, and interior detail
- Stairs
- Special floors and wainscot
- Plumbing and plumbing fixtures
- Heating and air-conditioning
- Electric wiring, lighting, and other electrical fixtures
- Decorating (painting and wallpaper)
- Finish hardware
- Special equipment (appliances)
- Porches or decks (framing, flooring, railings, pillars)
- Terraces
- Garages, garage doors, and carports
- Backfilling, grading, and drainage
- Other on-site improvements including termite treatment
- Landscaping and plants

- possibly describes the portions of the project for which each of the other parties is responsible
- authorizes the builder to purchase materials for the project
- identifies any unusual responsibilities that the buyer may request of the builder

The project shall consist of and the builder shall perform all of the work that is required by the contract documents, as follows:

(describe work)
____
____
____
____

The following items are specifically excluded from the terms of this contract, and the builder shall not be responsible for the same:

A. Interior papering, finishing of doors or any other decorating.
B. Installation or construction of walks, pavements, or curbing.
C. Ground fill, finish grading, seeding, planting shrubs, or landscaping.
D. Sewer or water permits.
E. Any item on the plans marked "Not in the contract" or "NIC"

Conformance with Plans and Specifications

Builders must explain to their buyers that the actual construction of their homes may deviate somewhat from the plans and the specifications or from models that the buyers may have seen. Furthermore builders may want to protect themselves from possible deviations from property lines, survey lines, and the like. These facts should be disclosed to the buyer in conversations at the beginning of the marketing and negotiation processes. These provisions should also be included in the contract and in the certificate of acceptance that is to be signed or initialed separately by the buyer as evidence that the builder disclosed the possibility of deviations or inconsistencies to the buyer. The buyer should initial each of the provisions in the sample language below.

> **Construction of the House**—The project shall be constructed to conform substantially with the plans and specifications that are set forth in this contract. The buyer acknowledges that in the course of construction of the house certain changes, deviations, or omissions may be necessary because of the requirements of governmental authorities, suggestions and design changes made by the architect or buyer, or particular conditions of the job. The buyer hereby (a) recognizes that minor changes may occur in the work and (b) agrees that as long as the project is substantially the same as described in the contract documents and within accepted industry tolerance, minor deviations will be accepted.

> **Furnishings and Models**—Furniture, wallcoverings, furnishings, and the like as shown in or about any model are for display purposes only and are not considered a part of such home for the purposes of this contract. Further, the location of wall switches, thermostats, plumbing, electrical outlets, and similar items may vary from home to home and may not be as shown in any model home. Any floor plans, sketches, or sales drawings are for display purposes only and may not be exactly duplicated.

Three • Contract Between Remodeler and Owner

As discussed in Chapter 2, a written contract records the exact terms of an agreement between parties. The agreement defines the scope and price of the product and also allocates the risks inherent in a particular transaction between the parties. The remodeler, like the new home builder, should have a written contract with the customer. This chapter should help remodelers and their attorneys write remodeling contracts that lessen the remodelers' risk.

No boilerplate contract clause is suitable for every situation. Therefore remodelers should critically examine every transaction into which they enter to foresee contingencies or events that could lessen the transaction's benefits to them. They should make sure that each contract is written to provide them adequate protection from events that may expose them to unintended liability. Many states regulate remodeling or home improvement contractors, dictating that certain information shall be included in any remodeling or home improvement contract. The regulation may also dictate that certain provisions appear in a certain style and size of type. Failure to comply with these laws may subject the remodeler to a fine or imprisonment. Accordingly remodelers are advised to consult their local laws when drafting their contracts (see sample state law in Figure 3-1).

Remodeling contracts and contracts for new construction share many of the same types of provisions, such as contract price and when it will be paid, a specific description of the work, the time for performance, change orders, and warranties. Differences exist between remodeling and homebuilding, however, and a well-drafted contract will reflect those differences.

The obvious difference is that a remodeler is working with an existing structure. As a result, removal of hazardous materials or waste may be a factor in remodeling, but it is not likely to be a factor in homebuilding. Similarly a remodeler may face the hazard of lead paint, but it is almost never a factor in homebuilding. Accordingly the remodeler's contract should expressly provide whether the owner or the remodeler is responsible for the removal of lead, asbestos, or other hazardous substances.

Matching materials is another potential source of trouble for remodelers, but not new home contractors. A remodeler may have to match existing materials that are no longer produced or materials that have faded because of age. To shape the owner's expectations about matching materials, remodelers should include specific language in their contracts to alert homeowners to the limitations of matching certain materials.

Another obvious difference between remodeling and homebuilding involves customer contact. Remodelers sell a service, whereas home builders, particularly speculation builders, also sell a product.[25]

Because remodelers are likely to be working in an owner's current home or place of business, they must be concerned with such items as—

- access to the work place
- working while people are living on the premises
- access to bathrooms and telephones
- removal of debris and daily clean-up
- protection of the owner's property
- ownership of salvage material

This chapter explains the purpose of various clauses and provides sample language that may be used in remodeling contracts between owners and remodelers.

These suggested provisions do not address every contingency nor do they apply to all remodeling agreements. This book also does not cover all of the provisions that a remodeler should consider for inclusion in a remodeling contract. However, if they are written correctly, these and other provisions may protect a remodeler against unnecessary liability.

Contracts are legal documents that greatly determine a remodeler's liability if a homeowner alleges that the remodeler failed to perform the contract obligations.

Figure 3-1. State Home Improvement Contractor Law; Contract Requirements

[Any changes to] California home improvement contracts that exceed $500 . . . must be presented in writing and include the following [items]:

- the name, address, and license number of the contractor
- the name and registration number of any salesperson who solicited or negotiated the contract
- the approximate dates when the work will begin and end
- a plan and scale drawing showing the shape, size, dimensions, and construction equipment specifications
- a description of the work to be done, descriptions of the materials to be used, a list of the equipment to be used or installed, and . . . compensation [agreed upon] for the work
- a schedule of payments and a statement that, upon satisfactory payment for any portion of the work performed, [before any further payment is made], the contractor shall . . . furnish the person contracting for the home improvement a full unconditional release from any claim or mechanic's lien, for that portion of the work for which the payment has been made
- a notice in close proximity to the signatures of the owner and contractor in at least 10-point type stating that the owner or tenant has the right to require the contractor to have a performance and payment bond
- a statement that no extra or change order work shall be required to be performed without written authorization of the person contracting for the construction of the home improvement
- notice that, if the contract provides for payment of a salesperson's commission out of the contract price, that payment shall be made on a percentage basis in proportion to the schedule of payments made to the contractor by the disbursing party in accordance with the schedule of payments
- a description of what constitutes substantial commencement of work pursuant to the contract
- a notice that failure by the contractor without lawful excuse to substantially commence work within 20 days from the approximate date specified in the contract for work to begin is a violation of the Contractors' State Licensing Law

A violation of this section by a contractor, his or her agent, or salesperson is a misdemeanor punishable by a fine of not less than one hundred dollars ($100) nor more than five thousand dollars ($5,000), imprisonment in the county jail not exceeding one year, or both that fine and imprisonment.

Therefore an attorney experienced in construction contract law should prepare (or at least review) any such documents before a remodeler signs them. To find such an attorney, the remodeler would contact the local NAHB Remodelors® Council, local home builders association, local branches of the Associated General Contractors or of the Associated Builders and Contractors, or State Bar Association. Sample language that appears in brackets presents an option for the reader. Depending on the particular transaction the reader may choose one or the other, both, or neither of the possible wordings presented.

Warning—The National Association of Home Builders has provided this guide and sample contract language merely to point out the types of provisions of which the remodeler should be aware. These suggested provisions should not be used without review by an attorney experienced in construction law.

This chapter addresses the major sources of potential liability that a remodeler may face. The written contract may encompass three documents: the agreement, the general conditions, and the plans and specifications. Often remodelers combine the agreement and general conditions in the same document. The contract also may include other documents and clauses. (In reading the sample language that appears in the section which follows, note that the items in brackets require a decision by the remodeler.)

The Agreement

The agreement is the document that the parties sign. The plans and specifications are named in the agreement as being part of the contract.

Caption

The caption is the heading or introductory part of a legal document. It should include the names of all persons listed as property owners. The full legal name of each party should appear in all documents in the transaction. If the owner is a corporation or other business entity, the documents should include the full name and type of business entity. The names of the parties should be exactly the same in the contract as in all other project documents. The caption also should give a legal description of the property (such as lot, block, subdivision).

> This contract specifies the terms between (owner's name), the customer, and (remodeler's name), the remodeler, to [remodel, renovate, rehabilitate, or build an addition to] the [home or other structure] on the property located at: (legal description of property).

The Contract Documents

The agreement that the parties sign contains such items as the names of the parties and other information specific to the project. But some terms are more fully described in other documents such as the plans and specifications. The remodeler's warranties and warranty limitations are also often drafted separately (see Chapter 6). If the written agreement specifically references these other documents, however, their requirements and limitations will constitute part of the contract (Figure 2-1). The agreement should specifically and accurately describe the referenced documents by date, number of pages, plan number, name, and any other appropriate information. The requirements contained in the incorporated documents should not conflict with each other nor with the terms of the agreement. Examples of contract documents appear in Figure 2-1.

> The terms of this agreement include the conditions of this agreement and, by reference, the provisions in the other documents specifically listed below. The terms of this agreement prevail over any conflicting provisions in the documents incorporated by reference.
>
> The remodeler shall perform all of the work that is required by this agreement and any documents incorporated by reference below.
>
> Except for written modifications signed by both parties subsequent to the execution of this agreement, the terms of the agreement are limited to the provisions contained in this agreement and the documents described as follows:
>
> - (title of document), dated the ________ day of ________, 19____, consisting of ____ pages;
> - (title of document), dated the ________ day of ________, 19____, consisting of ____ pages;
> - (title of document), dated the ________ day of ________, 19____, consisting of ____ pages;

Illustrative Case

In an action by a homeowner against a contractor for breach of a contract to remodel a house, the evidence supported the court's conclusion that the contractor was bound by the minimum building standards required by the homeowner's lender. The primary contract, dated and signed by the parties, stated that the contractor agreed to provide all materials and to perform all the labor shown on the work drawings and described in the specifications. The contract further provided that the contract incorporated the specifications, including the proposal signed by the parties. The proposal stated in part that all material was guaranteed to be as specified and that work would be performed in accordance with drawings and specifications submitted for the work. Finally the instructions incorporated in the description of the materials provided that "the specifications include this Description of Materials and the applicable Minimum Construction Requirements." The court's conclusion that the contractor was bound by the minimum requirements set by the lender was supported "by the fact that he signed a certification indicating that they would be followed and by the fact that their existence was recognized in the contract documents." Although some of the minimum requirements conflicted with the other specifications expressed in the contract, the contractor was the author of those terms and they were construed against him.[26]

Ownership of Plans and Specifications

If the owner provides the plans, the remodeler should specifically impose liability for defects in the plans on the owner. Remodelers who provide the plans, however, should expressly prohibit owners from giving or selling those plans to other potential owners or remodelers. The best way to ensure enforcement of this latter provision is to copyright the plans. Sample clauses are provided for both owners' and remodelers' plans.

> **Owners' Plans**—The remodeler agrees to perform the work in accordance with the building plans supplied by the owner (reference by title, author, and number of pages). Remodeler assumes no responsibility or liability for defects in the design or engineering of these plans. Only the owner will be liable for any damages caused by defective or negligently drawn building plans, including but not limited to additional costs caused by delay in substantial completion of the work, additional costs for materials necessitated by any changes, additional labor costs, and percentage profit on the additional work.
>
> **Remodelers' Plans**—The remodeler has provided the building plans to be used under this contract (reference by title, date, author, and number of pages). The owner of the property to be (remodeled, renovated, rehabilitated, or added to) has no ownership rights in the architectural plans used under this contract, and the owner will be liable to the builder in the amount of lost profits and all consequential damages for the reuse or resale of these plans.
>
> The remodeler makes no representations or warranties about the quality of these plans except those specifically provided in the limited warranty references in this contract.

Time of Commencement and Substantial Completion

The date work begins should be no earlier than the date of execution of the contract. Instead of a specific date, a flexible starting date may be established by using a notice to proceed; for example, "The work shall commence on the date stipulated in the notice to proceed." An alternative is to tie commencement to the date the building permit was issued. Some states and local jurisdictions require that the contract give start and completion dates. The remodeler should have evidence of the owner's financing before commencing work (see General Conditions). In addition, before beginning the project, the remodeler should require evidence of a valid title and copies of a current survey, deed restrictions, and easements, if any. These items may help protect the remodeler's lien rights if a lien must be filed on the property for lack of payment, or they may identify potential problems with the site that the remodeler can avoid.

The date of substantial completion of the work may be a specific date expressed as a number of days (preferably calendar days). The contract should make clear to the owner when substantial completion will occur (for example, the date that the certificate of occupancy is issued, actual occupancy or use by the owner, or some other identifiable point of time).

The contract should provide that the owner has a certain number of days to sign and return the contract. Additionally the contract should include a provision for extending the time of completion, as described in Liquidated Damages and Unavoidable Delay in this chapter.

> Assuming all conditions precedent are satisfied and weather permits, the work to be performed under this contract shall commence on the ____________ day of ________________, 19____, and be substantially completed no later than the ____________ day of ________________, 19____. These starting and completion dates are subject to the owner signing and returning the contract within 10 working days from receipt of the contract.

Price, Deposit, and Payment

This section of the contract specifies and defines the remodeler's compensation for his or her work. It also instructs the parties in the manner and time of payment. To achieve these goals, this section should—

- Include the cost of the work. Some remodelers charge owners a total fixed price for the project. Others bill on a cost-plus-fee or cost-plus-percentage basis. The cost-plus contract should specify the percentage or fee required, the guaranteed maximum price (if any), what constitutes costs (see section on costs), how the fixed fee is to be adjusted for change orders, the type of fee schedule, and all anticipated costs and fees, when applicable. Whatever the type of contract, it should specify (in both words and numbers) (a) the total price of the construction or sales transaction and (b) the amount of owner's deposit to be paid to the remodeler upon execution of the contract.

Fixed-Price Contract—The owner agrees to pay and the remodeler agrees to [remodel, renovate, rehabilitate, restore, or build an addition to] the building _______ for the consideration of ______________________ dollars ($__________), the total price. The owner will pay ____________________ dollars ($_____) to the remodeler as a deposit [if one is required] upon signing this contract.

Cost-Plus-Fee Contract—The owner agrees to pay and the remodeler agrees to construct the project for the consideration of the remodeler's actual costs and expenses (as defined in Section __________ of this contract, Costs to Be Reimbursed) plus _____ percent (_____%) of these costs and expenses. The owners will pay __________ dollars ($_____) as a deposit upon signing this contract [or on or before some date other than the date the contract is signed]. The owner will make progress payments [by the tenth (10th) day of each month] based upon applications for payments submitted by the remodeler. The owner will make final payment (including all costs and expenses plus the fee specified above) to the remodeler within ______________ (_______) days of substantial completion. Substantial completion has occurred when the owner can use the project for its intended purpose. The owner has ten (10) days from receipt of each bill to pay the remodeler. Payments due under the contract but not paid shall incur daily interest at the rate of ______________ percent (__________%) from the date payment is due.

- Establish due dates for payments. The due dates should be mutually acceptable to the remodeler and the owner and should reflect specific stages of construction. The contract may include an up-front payment payable upon the signing of the contract. Thereafter the payments could be triggered by the commencement of a particular phase of the project rather than upon the completion of a phase of the project. (Some remodelers use monthly bills to show the progress of the job). Regardless of the plan, final payment should be due upon substantial completion and not upon final completion.
- State how much time the owner has to make the payment after receiving the request for payment. For example, the contract might state, "The owner has ten (10) days from the statement date to pay the remodeler."
- Require interest for late payments. For example, "Payments due under the contract but not paid shall incur daily interest at the rate of ________ percent (____%) from the date payment is due."

Instead of charging interest on late payments, the contract might provide for a late payment fee similar to liquidated damages. Such a late fee might be included in a cost-plus-fee contract as well as in a fixed-cost contract using this alternate language:

> The owner hereby acknowledges that late payment by the owner to the remodeler of progress payments and final payment may cause the remodeler to incur costs not contemplated elsewhere in this contract, the exact amount of which will be difficult to ascertain. Accordingly any sum due the remodeler under this contract shall be paid within ten (10) days of written request. If full payment is not received within ten (10) days, the buyer shall pay ten percent (10)% of the total amount overdue as a late charge. The parties agree that such late charge represents a fair and reasonable estimate of the costs the remodeler will incur by reason of late payment by the owner.

- Specify whether retainage is allowed. If it is allowed specify when it will be disbursed.
- Require evidence of financing. The remodeler should be allowed to verify that the owner has satisfactory financing, including an allowance for subsequent change orders. The owner can be made to produce evidence of financing before the remodeler is obligated to start work.
- Satisfy any special state or local requirements. Some states require that the contract specifically identify certain expenses.
- Specify that work covered by change order falls within the general scope of the work contemplated by the contract, and specify a percentage of profit for additional work. Change orders for deductions from the work usually involve no reduction of profit, and this should be clearly stated.

> If the buyer requires an addition to the scope of the work, the builder shall account for the cost of

these additions separately and shall be reimbursed __________ percent (_____%) of the cost of the changes.

Costs to Be Reimbursed

This section is applicable to cost-plus-percentage or cost-plus-fee contracts. This list of items suggests some costs that may need to be reimbursed in a remodeling project, but it may not include all costs to be reimbursed on every job. The remodeler should include all relevant items, especially every contingency negotiated with the owner. Remodelers should check this section against their general ledgers and other accounting practices to make sure all appropriate costs are included.

Many contracts will also include a section regarding costs not to be reimbursed. These provisions should be carefully reviewed because by definition they exclude some of the remodeler's actual costs. These items would be paid out of the percentage or fee added to the defined costs and, therefore, cut into a remodeler's profit.

The term *cost of work* means costs necessarily incurred in the remodeler's proper performance of the work contained in the contract documents, including the following items:

- Wages, benefits, costs of contributions and assessments for workers' compensation, unemployment compensation, Social Security, taxes, or any other costs incurred by the remodeler for labor during performance of this contract.
- Salaries for remodeler's personnel (including production supervisor or manager, where applicable) to the extent their time is spent on work helping to complete this contract. This payment of salaries includes (but is not limited to) work performed at shops, on the construction site, or in transporting materials or equipment.
- The remodeler's and the remodeler's agents' reasonable travel and subsistence expenses incurred toward the completion of this contract.
- Payments made by the remodeler to subcontractors for work performed pursuant to subcontracts under this agreement.
- Cost of all materials, supplies, and equipment incorporated in the work and consumed in the performance of the work; cost less salvage value of such items used, but not consumed, that remain the property of the remodeler, including all costs of installing, repairing and replacing, removing, transporting, and delivering the machinery and equipment.
- Rental charges consistent with those prevailing in the area for machinery and equipment used at the construction site, whether rented from the remodeler or others, including all costs of installing, repairing and replacing, removing, transporting, and delivering the machinery and equipment.
- Sales, use, excise, or any other taxes related to the work imposed by governmental authorities.
- Permit fees, charges, inspection costs, or any other fees related to the work imposed by governmental authorities.
- Royalties, damages for infringement of patents, costs of defending related lawsuits, and deposits lost for causes other than the remodeler's negligence.
- Expenses for telephone calls, telegrams, postage, delivery fees, stationery, and other similar petty cash items related to work.
- Costs incurred for security at the job site.
- Costs incurred because of any emergency affecting the safety of persons or property.
- All landscaping and backfilling necessary under the contract documents, including the cost of all grading, removing or planting of trees, removing snow, frost breaking, pumping water, excavating and related work, delivering the necessary materials to the site, and hauling away excess fill and material or trash and debris.
- Remodeler's risk or other insurance, soil fees and civil engineering fees, performance bonds, and labor and suppliers bonds in an amount equal to one hundred percent (100%) of the maximum costs.
- Differing site conditions, as provided for in Section _____ of this contract.
- Any other costs incurred in the work that are within the scope of the work as defined in the contract documents, or any other overhead expenses.

Illustrative Case

In an action to foreclose a lien on the owners' residential property for work done and materials furnished in remodeling and redecorating under a cost-plus contract, the court held that the contractor's charges for certain work and materials were excessive. The court's decision was based in part on a Mississippi case from which it quoted as follows:

"The rules of law controlling 'cost plus' contracts are well established. Upon reason and authority, where a person agrees to do work for another upon a cost plus basis, it is his duty to keep accurate and correct accounts of all materials used and labor performed, with the names of the materialmen and laborers, so that the owner may check ... [on] the same. He must use the same skill and ability as is used in contract work for a gross sum. If the aggregate cost upon the face of the account is so excessive and unreasonable as to suggest gross negligence or fraud, the law would impose upon the contractor the duty of establishing the bona fides of his performance of the work. The contractor does not have the right to expend any amount of money he may see fit upon the work, regardless of the propriety, necessity, or honesty of the expenditure, and then compel repayment by the other party, who has confided in his integrity, ability, and industry....

"In an action upon his contract for payment, the contractor must show that the moneys which he claims to have expended were necessarily paid for materials and work upon the job and if the contractor fails to do this he should only be allowed the reasonable cost and his percentage...."[27]

Draw Schedule and Application for Progress Payment

The remodeler and the owner may negotiate any form of payment schedule that is mutually agreeable. In the absence of a payment schedule, the remodeler is not entitled to receive payment until the work is completed. Typically, if the owner is financing the project, the method of payment is through "progress payments" from the owner to the remodeler. One alternative allows the remodeler to establish a draw schedule with the homeowner and construction lender. With a draw schedule a specified amount (for example, 10 percent) is sometimes paid up front, the remodeler can withdraw from the account at successive phases during construction, and the lender may hold a retainage until the structure is accepted by the owner.

A draw schedule should require the owner to make or authorize each payment to the remodeler within a specified number of days after the remodeler's application for payment. The remodeler uses the application for payment document, along with various supporting documents, to establish the remodeler's right to payment. Before requesting final payment, the remodeler should ensure that all claims have been settled and should obtain signed confirmations from the owner that, based on the final inspection, the contract requirements have been fulfilled.

Alternatively the contract may provide for inspection by the lender or another third party at certain stages of the construction, with payment due directly from the lender upon satisfactory inspection. The parties should specify who will schedule the inspections and who will pay the inspector. In any case the contract should specify whether the progress payments provision includes payment not only for the work in place, but also for materials and equipment suitably stored at the site or at an off-site location agreed upon by the owner.[28]

Such a provision might give the owner the right to withhold progress payments otherwise due to the remodeler under certain circumstances enumerated in the contract, for example, if the remodeler fails to pay his or her subcontractors or suppliers. Similarly this provision might provide that the remodeler can stop work until he or she is paid, and if payment is not made within a specified number of days after it is due, the remodeler also may terminate the contract.

The following sample language provides for two methods of payment.

Regular Periodic Payment—The remodeler shall prepare [weekly, biweekly, or monthly] itemized statements for the owner that specify all costs incurred by the remodeler in furtherance of performance of the terms of this contract. The owner will place adequate funds in an escrow account from which the remodeler may draw to pay for the cost of the work. Each draw will be based on an application for payment submitted to the owner by the remodeler. The owner will make funds available to the remodeler within ________ (____) days of receipt of the statement (less _______ percent [____%] retainage) as provided for in the Notices provision of this contract.

The owner shall make final payment to the remodeler, including the entire unpaid balance of the cost of the work and all other compensation due to be paid to the remodeler, as defined in Section ______, Costs to Be Reimbursed, of this contract [at final settlement or within ________ ( ) days after substantial completion of the work].

Payments Tied to Construction Activity—The remodeler shall prepare [weekly, biweekly, or monthly] itemized statements for the owner that specify all costs incurred by the remodeler in furtherance of performance of the terms of this contract.

The owner will place adequate funds in an escrow account from which payments are made for the cost of the work. Each draw will be based on beginning various phases of the work as described below. The owner will make funds available to the remodeler within __________ (___) days of receipt of the notice (less _____ percent [____%] retainage) as provided for in the Notices provision of this contract.

The remodeler agrees to furnish materials and labor in accordance with the provisions of this contract for __________ dollars ($_______) payable in the following stages:

1. $_____ upon signing the contract
2. $_____ upon start of _______________ ___________________________________
3. $_____ upon start of _______________

[Depending on the job, it might have 2 to 10 stages.]

The owner shall make final payment to the remodeler, including the entire unpaid balance of the cost of the work and all other compensation due to the builder, as defined in Sections ____, Price and Deposit, and ____, Costs to Be Reimbursed, of this contract upon substantial completion of the work. This project shall be substantially complete upon the issuance of a certificate of occupancy. The owner may retain ___________ dollars ($_______) for ________ days after substantial completion to assure—

- final issuance of permits required of the remodeler under the contract
- correction of defects
- system performance
- passage of inspection

Delays and Early Completion

Liquidated Damages and Unavoidable Delay

A liquidated damages clause is a provision for the payment of predetermined money damages in the event of a breach of contract, such as inexcusable late completion of the work or inexcusable late payment of a draw. In this context the word *liquidated* means *determined* or *settled* and is used to indicate damages that are agreed to or settled in advance.[29]

Contracts with subcontractors and suppliers should include a provision stating that they have read and are bound by the remodeler's agreement with the owner (see Chapter 8). This statement will put the subcontractors and suppliers on notice that the remodeler may be liable to the owner in the event of a delay in construction, and if the delay is caused by a subcontractor or supplier, that firm may be liable to the remodeler for damages.

The liquidated damages clause should not impose a penalty on the party who fails to perform, but it should reasonably approximate the actual losses that may occur because of the nonperformance. If the contract allows liquidated damages against the remodeler for work not completed on time, the liquidated damages clause should specify that the remodeler is not responsible for delays beyond his or her control. If the contract allows the remodeler to extend the time for performance, the remodeler should notify the owner in writing of his or her intention to do so. The remodeler should not wait until after the time for performance has passed to explain to the owner that he or she was delayed, thereby necessitating an extension of time to perform. Estimating actual losses an owner would suffer can be difficult, and if the owner cannot establish actual damages a liquidated damages clause may work against the interests of the remodeler. However a homeowner may suggest the use of a liquidated damages clause. Therefore the remodeler should be prepared to address this issue. In addition a liquidated damages clause may actually benefit the remodeler by limiting the damages to the amount agreed upon if the contract specifies that the owner's liquidated damages remedy is exclusive of all other remedies.

In the event that the work to be performed under this agreement is not substantially completed by the completion date, the remodeler will compensate the owner in the amount of ____________ dollars ($_______) for each day of the week (including weekends) of inexcusable delay until the work is substantially completed. If the remodeler's failure to perform the terms of this contract in full does not result in any additional expenses to the owner, the remodeler's damages shall be limited to the cost of completion of performance of the terms of this contract. The owner shall not recover any compensation from the remodeler for delay caused by differing site conditions, as defined and provided for in Section ____ of this contract.

The remodeler will use his or her best efforts to complete construction of the project before the completion date. However if reasons beyond the remodeler's control cause an unavoidable delay in the progress of construction (including, but not limited to, such factors as the unavailability of materials, inclement weather, strikes, changes in governmental regulation, acts of governmental agencies or their employees, acts of God, work changes made by the owner, or the failure of the architect or the owner to cooperate), the remodeler may, in his or her sole discretion, extend the date of completion for a period equal to the time of the delays.

This liquidated damages remedy is exclusive of all other legal and equitable remedies.

Illustrative Case

James Brink, as president of Audubon Builders, Inc., entered into a contract for the building of an addition to the home of the Hanrahans on November 21, 1987. Mrs. Hanrahan's mother was to reside in the addition. The contract was signed with an addendum in James Brink's writing which stated: "All work to be completed by 2-15-88 or contractor agrees to $150/day compensation." The project was ultimately completed on December 22, 1988. The Hanrahans sued Audubon and Brink alleging that certain work had not been completed in a proper and workmanlike manner, and alleging that they were entitled to compensation in the amount of $150 per day because of the delay in the completion of the contract. The trial court awarded the Hanrahans $150 a day in liquidated damages from February 15, 1988, through December 22, 1988, and Audubon and Brink appealed. The appellate court reversed the trial court opinion because that court failed to consider whether the $150-a-day liquidated damages provision was a reasonable approximation of probable damages or that actual damages would be impossible or difficult to calculate. The court concluded that the provision was agreed to by the parties as a penalty to ensure performance and was not reasonably related to the actual damages caused by the delay in performance.[30]

Bonus for Early Completion

As a corollary to the liquidated damages clause, remodelers often include a bonus provision that entitles them to a per diem sum of money for finishing early.

If the remodeler substantially completes the work before the substantial completion date in Section ____ of this contract, the owner will pay the remodeler the sum of ________________ dollars ($________) for each day of the week, including weekends, between the date when the work is substantially completed and the substantial completion date set by the contract.

Notices and Records

The contract is written to minimize disputes about the terms of the agreement. Likewise all notices provided for under the contract should be written. The remodeler should keep all records related to the project for at least the duration of the project. Ideally the remodeler should store the records until all applicable statutes of limitations have expired. The statute of limitations is the period of time within which an owner must file a lawsuit. That period can run for many years, and under certain circumstances the courts may extend the limits indefinitely.

Any notice required or permitted to be delivered under this contract should be mailed to the parties at the following addresses:

____________________	____________________
(remodeler)	(owner)
____________________	____________________
(street address)	(street address)
____________________	____________________
(city, state, zip)	(city, state, zip)

For purposes of this contract, notice is received when sent by certified mail, postage prepaid, return receipt requested via the United States Postal Service.

Effective Date And Signatures

Ideally all parties to a contract sign the document at the same time. If that is not possible, however, this provision assures that the remodeler is not bound by the terms of the contract until the remodeler (or an authorized representative) has signed the contract.

If any changes are made to the contract after it is printed or typed, both parties must initial each revision for it to be valid. Of course a change order should cover any changes to the work after the contract is signed in accordance with the change order requirements of the section on Changes in Work. Any changes to the contract after it is signed require a written amendment according to the section on Entire Agreement and Severability.

The owner and the remodeler should sign at least three original contract documents. (Computer-generated hard copies are useful for this purpose.) The parties should execute the contract in their proper capacities, as individual owner or as partners, officers, or directors of a partnership or corporation that is a party to the contract. The parties should use their complete legal names on all contract documents and any subsequent change orders or contract amendments. Once both parties have signed the document, they should complete and initial the date-of-contract provision.

This contract has no force or effect and will not be binding upon the remodeler until it is accepted and executed by the owner and countersigned by the remodeler. This contract shall become effective on the date designated below or the day it is executed by both parties, whichever is later.

We, the undersigned, have read and understand and agree to each of the provisions of this contract and hereby acknowledge receipt of a copy of this contract.

______________________	______________________
(owner's signature)	(name of remodeler, corporate name where applicable)
______________________	By ______________________
	(authorized signature)
Date ______________________	Title ______________________
______________________	Date ______________________

This contract is dated, and becomes effective:

______________________	______________________
(month, day, year)	
______________________	______________________
(buyer's initials)	(builder's initials)

Illustrative Case

When a remodeler's corporate existence did not begin until some 14 days after entering into a contract with the owners for the construction of an addition to their home, the court properly found the remodelers individually liable for breach of the contract.[31]

General Conditions

Every construction contract has terms and conditions that apply to all contracts and that are not specific to a single project.

General conditions are sometimes included on the same page as the agreement. Some builders include the agreement and the general conditions in one document; others separate the two.

Obviously the contract cannot cover every contingency. The more time the parties spend identifying their respective obligations before the project begins, however, the fewer surprises will occur later.

Financing and Other Contingencies

If he or she has not already done so, the owner should agree to apply for a loan within a certain period after execution of the contract. If the owner does not receive a loan, the contract is null and void. The remodeler should not begin work under the contract until the owner has obtained the necessary financing.

An owner who does not need financing should place funds equal to the total price of the agreement in an escrow account. Similarly, if the owner needs only partial financing, the balance of the funds due should be placed in escrow.

The owner represents that he or she has arranged sufficient financing to comply with this agreement. Before commencing work under this agreement or at any time during the progress of the work, the remodeler may request evidence in writing (acceptable to the remodeler) of financing for the work. Failure of the owner to produce the requested information within seven (7) calendar days of the written request will constitute a breach of contract by the owner, and the remodeler may suspend the work. If the owner fails to provide the requested information within an additional seven (7) calendar days thereafter, the remodeler may terminate this agreement. Should the remodeler suspend or terminate this agreement, he or she shall be entitled to collect payment for materials and labor expended on or for the project, along with a reasonable profit and overhead.

Any other contingencies that the parties contemplate should be specifically identified. The contract should specify the period in which those events or requirements must occur, what will happen to the agreement if the conditions are not fulfilled, and how the deposit money will be divided.

Allowances

The total contract price shall include any allowances mentioned in the contract documents. Allowances are frequently included when an item of work is not sufficiently detailed in the contract documents to enable the remodeler to determine the final cost of the item. Their inclusion in the contract establishes that the item is within the scope of the remodeler's scope of work subject to final price adjustment.[32]

Allowance items may be of two types: (a) items that the owner still needs to select, such as wallpaper, cabinets, and the like, and (b) items of work for which the actual cost cannot be determined until actual conditions are verified or until additional information regarding the scope of the work is received. In the former situation the remodeler should closely monitor the owner's spending habits during construction and periodically notify the owners of the status of the allowance. In the latter situation the remodeler must be able to justify the original estimate because, if the actual cost greatly exceeds the allowance, the remodeler may have to deal with an angry customer. If the allowance was grossly underestimated, a court may allow the remodeler to recover only his reasonable costs.

The prudent remodeler will provide in this clause or in the separate Allowance Schedule that if costs are more than or less than allowances, the contract price shall be adjusted accordingly by change order, and that the change order shall reflect the difference between actual costs and the allowance. The remodeler should make sure that the owner understands that, if the actual cost exceeds the allowance, the owner is responsible for paying the difference. The remodeler shall supply items for which allowances are provided within the amounts specified. Those amounts must cover the applicable taxes as well the builder's cost for materials and equipment delivered to the site. Trade discounts will be passed on to the owner. The allowance item should clearly state whether the allowance is for material only, installation only, or material and installation.

To avoid delaying the work, the owner must promptly select all materials and equipment for which allowances are provided. So long as the remodeler has no reasonable objections to the vendors selected, the owner may specify the vendors for these items.

Illustrative Cases

Case 1—The owners were not entitled to a credit for the amounts by which they exceeded the stated allowance for items such as carpets, cabinets, and the heating system where the evidence established that the owners understood that they would have to pay the difference if they exceeded the stated allowance.[33]

Case 2—The court ruled that a general contractor-landlord could not recover all its expenses incurred to "finish out" space occupied by tenants under a lease, although the tenants exceeded their allowance by $16,270.81, because the landlord failed to notify the inexperienced tenants, as the work progressed, how much of the allowance they used. The court held that the landlord was not entitled to recover expenses in excess of the contract allowance from its tenant because "dealing with neophytes, they had an obligation of fair dealing, which required some notification to defendants of the status of the allowance at a meaningful time." The court noted that the tenants relied on the landlord to do the job within the allowance. However they did not communicate that expectation explicitly to the landlord, and, despite its knowledge that tenants often have unrealistic expectations, the landlord provided only a subtle warning. The court observed that the case presented the classical perils of a failure to communicate.[34]

Permits, Licenses, and Other Approvals

Typically the remodeler is responsible for obtaining all permits, licenses, fees, and approvals associated with the construction and occupancy of the project. A remodeler must be careful, however, about agreeing to obtain all permits. The remodeler is probably better off identifying in the contract the various permits, licenses, fees, and approvals for which he or she will be responsible. The remodeler who agrees to obtain all permits, for example, probably contemplates obtaining the local building permit and the certificate of occupancy permit, but not a wetlands permit or a state water quality certification. However if the remodeler agreed to obtain all permits, and the owner's lot contains wetlands, the remodeler may be required to obtain a wetlands permit before he or she may continue with the project. The permitting process can be lengthy (in some cases more than a year) and expensive. In addition, if the remodeler agrees to obtain the building permit, the contract should provide for the contingency that a permit will not be issued.

If the owner lives in a development that requires approval of plans and specifications, materials, and colors by an architectural review committee, these approvals should be obtained by the owner.

> The remodeler shall obtain and pay for all local building and construction permits, licenses, governmental charges and inspection fees, and all other approvals necessary for work, occupancy of permanent structures, or changes in existing structures that are applicable at the time this contract is signed, excluding variances and other changes in zoning. The owner is responsible for negotiating for contested permits, licenses, and other approvals. In the event the remodeler cannot obtain the building permit within __________ (_____) days of the effective date of this agreement, the remodeler may cancel this agreement, thereby relieving the parties of any further obligations under this agreement.
>
> If required, work is to be undertaken under the following permits that will be obtained and paid for by the owner or remodeler as designated below:

Insurance and Risk Management

The purpose of risk management is to reduce or eliminate the risks associated with financial loss. Insurance is a key component of risk management.

Many types of insurance coverage are available to remodelers, and an agreement to buy insurance is often written into remodeling contracts. The contract will generally require both parties to buy and maintain insurance for specified injuries or risks and in specified dollar amounts. The contract should clearly state the insurable interest of each party and establish who owns or is responsible for what property, during what period of time, and the type and amount of coverage required.

The need for insurance and the exclusions to insurance policies vary according to the type of construction involved. Some common types of insurance include the following:

- comprehensive general liability
- builder's risk
- workers' compensation
- automobile liability
- property insurance
- professional liability
- completed operations
- umbrella liability
- contractor's equipment insurance

The contract should require the owner to raise his or her coverage to include the added value of the remodeling. The contract also should require the owner to list the remodeler and his or her subcontractors as additional insureds on any existing insurance policy. The contract also should require the owner to obtain insurance against loss resulting from injury to third parties or their property. Third parties could include persons not parties to the contract or the owner's employees or invitees. Remodelers should consult a construction insurance expert to determine their insurance needs, and an attorney for specific contract language.

Risk of Loss

Remodelers obtain insurance to cover numerous risks, but one risk that goes to heart of the transaction is the risk of loss caused by accidental destruction of or damage to the house during remodeling. In addition to obtaining insurance to cover such a loss, the parties to a contract should discuss their respective rights and obligations if such a contingency occurs. Destruction of the building excuses the parties from further performance under the contract and entitles the remodeler to recover the value of work and materials furnished before the destructive event. Despite the general rules the parties would be wise to expressly provide who must bear the loss under the contract if the work is accidently destroyed or damaged.

Illustrative Case

The Kaufmans engaged Gray to make certain renovations in their home for a total price of $5,000, payable one-third when work commenced, one-third when the work was half completed, and the balance upon completion. After the first two payments had been made, but before completion of the work, a fire at the house destroyed or damaged much of the work that had been done. Gray sued for the full balance remaining unpaid under the contract. The court held that Gray was not entitled to recover the balance of the contract because he had not substantially completed performance before the fire. The court held that where a contractor engaged in the repair of a building is prevented by fire on the premises from completing his contract, recovery is based on the value of work done and materials furnished by the contractor in the performance of his contract before the destruction of the property made full performance impossible.[35]

Differing Site Conditions

Unexpected site conditions, such as soil of inadequate bearing capacity; rock or other material not removable by ordinary hand tools; hidden pipes, sprinkler lines, and water and sewage disposal systems; or an unexpectedly high water table can be expensive to work around. Accordingly a prudent remodeler would include in the contract language contemplating the effect of unexpected or concealed conditions that may exist where the remodeler is building according to plans and specifications produced by the owner.

This provision protects remodelers and owners if such conditions are encountered. It protects remodelers because it eliminates the risk of increased cost from unexpected latent conditions. Without such a clause, a remodeler who agrees to construct something under a contract is not entitled to additional compensation merely because unforeseen difficulties are encountered. This provision also protects the owner because it obligates the owner to pay only for differing site conditions actually encountered. Remodelers who are responsible for differing site conditions usually increase their estimates to compensate for them. If the site conditions later prove to be normal, the owner will have paid for costs that never materialized. Under normal site conditions, the differing site conditions clause benefits the owner because it eliminates a windfall to the remodeler.[36]

The contract should include a clear definition of differing site conditions and provide for notification, work changes, or contract modification procedures upon discovery of such a problem. The definition should be broad enough to include unknown conditions in an existing structure in addition to concealed conditions below the surface of the ground.

Generally remodelers face two types of differing site conditions: (a) conditions that vary from conditions indicated by the contract and (b) unusual and unknown physical conditions that differ materially from those generally recognized as inherent in the work of the type covered by the contract.

An alternative to the clause provided below could require the owner to pay the extra costs incurred, plus profit at a specified percentage of the cost. The owner or remodeler may also have the option of canceling the contract. These suggestions would replace the equitable adjustment language below.

Under federal government contract law, the term *equitable adjustment* has an exact meaning. Outside this special area of law, however, the definition of equitable adjustment can be confusing. Remodelers' attorneys should check the case law precedents in their states to see how the term is treated by the courts and should use it only where the term has an identified meaning.

> A differing site condition is a physical characteristic of the property that materially changes the construction techniques from those reasonably expected at the time of the contract. Examples of differing site conditions are subsurface or latent physical conditions at the site materially different from those indicated in this contract or unknown physical conditions of an unusual nature (that are not reasonably foreseeable) located on the building site.
>
> Before disturbing any differing site condition, the remodeler shall notify the owner of such a condition except in the case of eminent danger to persons or property. The owner shall investigate the condition within five (5) business days. If the owner and the remodeler agree that the differing site conditions will cause an increase in (a) the remodeler's cost of performance of any part of the work under this contract or (b) the time required for that work, the issue will be resolved as follows:

A. If the total contract price will increase by more than ________ percent (____%) [for example, 10 percent], the owner may terminate the contract upon paying the remodeler for all costs expended to date and for the remodeler's full profits as provided in Section ______, Price, Deposit, and Payments, of this contract; or,

B. The remodeler and the owner shall execute a written specific cost adjustment to this contract, including any adjustments in the time for performance necessitated by the differing site conditions.

The owner's failure to investigate the condition will confer authority upon the remodeler to complete construction of the project according to the terms in Section ____, Liquidated Damages and Unavoidable Delay, of this contract and the builder shall further be entitled to [an equitable adjustment or payment of any increased costs necessitated by the differing site condition].

If the parties cannot agree on the existence or consequences of a differing site condition, the terms of this provision shall be arbitrated as provided for in Section ____, Arbitration, of this contract.

Architects and Designers

Before they talk to a remodeler some owners consult an architect or designer who continues to participate in the project. In such a case the contract should name the architect or designer and designate whether or not he or she is the owner's agent.

Warning—A remodeler should not rely on an architect or designer unless that person has written authorization to act on behalf of the owner. If the contract provides such authorization, it also should specifically provide that the remodeler will not be held liable for any actions made in reliance on recommendations of the architect or designer.

The architect or designer typically is responsible for—

- inspection or observation
- processing change orders
- overseeing the payment process
- interpreting plans and specifications with regard to aesthetic considerations

Of course the architect or designer cannot change the obligations of the parties as spelled out in the contract.

Sometimes the architect or designer has a role in disputes. The parties could agree that before a dispute can be submitted for arbitration or litigation, for instance, it could be submitted to the architect for consideration as a mediator.

This contract specifies the terms between (owner's name), the owner, and (remodeler's name), the remodeler, to [remodel, renovate, rehabilitate, or build, an addition to] a home on the property located at: (legal description of property).

[Name of architect and/or designer] is the owner's agent, and the remodeler may rely on representations, statements, revisions, and approvals made by [architect and/or designer] related to the performance of the terms of this contract. The remodeler will not be held liable for any actions made in reliance on recommendations of [architect and/or designer].

Changes in Work and Change Orders

The owner does not have the automatic right to order changes in the work unless the contract confers that right. However most contracts include a change order clause to give the owner that right. This section sets up the procedure for writing change orders and explains how the cost will be determined, including overhead and profit. The actual change order is a separate document (Figure 2-3).

Often, after the contract is signed and work commences, the owner will request changes in the design or materials used. Before beginning any new work created by an owner's request for a change, the remodeler should require that the owner sign and pay for a written change order. A change order is an agreement specifying (a) revisions in the work and the price, (b) a revised payment schedule, if necessary, and (c) a new date for substantial completion, if necessary.

If the owner and the remodeler can agree to the cost of the change, a change order should be prepared and signed by both. If they cannot agree on the price, the contract could include a provision for the work to be done with the remodeler to be paid on a cost-plus basis. If more than one owner is involved, getting both owners to sign a change order may be time-consuming. To expedite the work the contract might provide that either owner may sign a change order as the agent for the other and that the signature of one owner is binding on the other owner.

A change order should address all changes that affect the scope of the work, the contract price, and the time for performance, not just changes in the work requested by the owner. Thus additional work to be performed under a differing site or concealed conditions clause should be handled by a change order. Similarly any increase in the scope of the work that is required by the building or planning department and that is not the remodeler's fault should be handled by a change order. The contract should expressly provide that such an increase will be treated as extra work and that the extra work will be executed upon written change order. Generally payment for each change order is due (a) when the work is performed, (b) upon completion of the change order, or (c) before the next draw. A remodeler who performs work without obtaining written change orders and who then presents the owner with a large bill for that work at closing is asking for trouble.

> The owner may request changes in the work within reason. Upon written directive by the owner, the remodeler will make changes, additions, or alterations. If the owner and the remodeler agree on the cost of the modifications, they shall execute a written change order describing the changes to be made and any changes to the contract price or completion date (Figure 2-3). Change orders shall be signed by all parties and become part of this contract, and the owner agrees to pay the remodeler for changes [upon completion of the work performed under the change order or by the next draw]. The buyers agree that either of them may sign the change order and that the signature of one is binding on the other.
>
> If the parties cannot agree on a fixed cost of the change, the remodeler will make the changes and will receive from the owner an equitable adjustment to include the costs of labor, materials, equipment, and supervision plus ________________ percent (_____%) of such costs. The owner agrees to make requests concerning any changes, additions, or alterations in the work in writing directly to the remodeler named in this contract and not to the workers, including subcontractors and subcontractors' workers on the job.

Illustrative Case

A contractor entered into a written contract to remodel and renovate a home and later submitted a revised proposal for more money as the result of extra items requested by the owners. The contractor could not recover the additional costs because a Connecticut statute required that home improvement contracts must be in writing. Relying on the legislative history of the statute, the court refused to find an exception to the writing requirement for contracts that had been partly or fully performed by the contractor.[37]

Mechanic's Liens

Mechanic's lien laws (construction lien laws in some states), ensure that participants in the construction process get paid for their work by granting them a specific interest in real property that has been improved by their labor or materials. All 50 states, the District of Columbia, and Puerto Rico have such laws. The requirements of these laws vary considerably from state to state. To benefit from or comply with a lien law, a remodeler, subcontractor, or supplier must strictly follow the procedures and requirements of the law.

The claimant (the person claiming the lien) must provide one or more forms of notice before the lien can be effective. This process is often referred to as "perfecting" the lien (language that is commonly used in the mechanic's lien statute). To perfect the lien, often the claimant must file a formal Claim of Lien within a fixed period of time after completion of the contract or the last furnishing of services or materials. When the lien has been perfected, it may be enforced in a lawsuit to compel the sale of the property. The time for enforcing the mechanic's lien varies from state to state, but seldom exceeds a year after the date on which the lien was filed.

A majority of states allow a remodeler to waive his or her lien rights by signing a lien waiver before beginning work for the owner. The lien waiver clause typically provides that the remodeler will not file any liens against the property for labor, material, or equipment furnished under the contract. A lien waiver clause offers little, if any, benefit to a remodeler. Homeowners find such a provision attractive because it ensures that the property will remain unencumbered by liens.

Obviously remodelers should be wary of signing any form that has the effect of waiving their lien rights before receipt of payment; moreover they should be on the lookout for typical mechanic's lien wording because the owner's attorney or the lender may suggest adding it.

> The remodeler unconditionally waives, releases, and relinquishes all right to file or maintain any mechanic's lien or other claim in the nature of a lien against the real property or building improved under this contract for (a) labor, material, or equipment that is furnished or that may be furnished by the remodeler under this contract or (b) any claim for extras, change order work, or increased costs.

Inspection, Acceptance, and Possession

Inspections allow owners to identify, and give remodelers the opportunity to correct defects in materials or workmanship. Because the owner or the owner's agent makes periodic inspections, the remodeler can document an owner's satisfaction with the completed work. The remodeler should keep a record of each inspection for at least the duration of any statute of limitations or repose (Figures 7-1 and 7-2).

Statutes of repose and statutes of limitation limit the time within which an injured party may sue the person who caused the injury. The statutes differ with regard to when the "clock begins to tick." A statute of repose begins to run at an arbitrary point in time, such as when the building is substantially completed and usually before the person suffers a loss. Conversely the time allowed by the statute of limitations begins to run when the injured party's cause of action arises; that is, when he or she discovers or reasonably could have discovered the problem. In practical terms, from the remodeler's perspective, the difference between the two is that without the statute of repose the remodeler could be liable for the entire life of the structure.

The contract should explain what steps the parties must take before the owner may use the space involved in the project. For example, the contract might include provisions for inspection, formal acceptance, and final payment.

> From time to time and upon substantial completion, the owner or the owner's designated representative shall inspect the house in the remodeler's or remodeler's representative's presence. At the time of the final inspection for substantial completion, the owner will give to the remodeler a signed punchlist that identifies any alleged deficiencies in workmanship or materials.
>
> The remodeler shall correct any items on the owner's punchlist that are, in the good faith judgment of the remodeler, deficient in workmanship and/or materials according to the contract. If the contract is silent on the standards, the remodeler must meet the standards of construction relevant to the community or area in which the project is located. The remodeler shall correct those defects at his or her cost within a reasonable period of time.

Representations and Warranty

The remodeler should discuss the limited warranty with the owner at the beginning of the negotiations (see Chapter 6). Most courts will uphold limitations on or exclusions from a remodeler's warranty if they are considered part of the contract. That is, if warranty limitations and exclusions are included in the give and take between the remodeler and the owner when they are forming their agreement, the courts will usually enforce these limitations and exclusions.

But many courts also require that, to be enforceable, the disclaimer of implied or other warranties must be conspicuously displayed in the contract. The disclaimer should appear either on the first page of the contract or just above the owner's signature. Putting the disclaimer in bold type also may help. Finally remodelers should require the owners to initial this section separately. In any event the remodeler should be sure that any disclaimer satisfies any applicable state or local laws. For example, in Indiana a remodeler may disclaim all implied warranties only if the following minimum conditions are met:

- The warranties defined in the Indiana statute are expressly provided for in the home improvement contract between a remodeler and an owner.
- The performance of the warranty obligations is guaranteed by an insurance policy in an amount equal to the contract price made under the home improvement contract.
- The remodeler carries completed operations products liability insurance covering the remodeler's liability for reasonably foreseeable consequential damages arising from a defect covered by the warranties provided by the remodeler.

Moreover the disclaimer shall be printed in at least 10-point boldface type, and the owner must acknowledge the disclaimer of implied warranties by signing a separate one-page notice attached to the home improvement contract. (Indiana Statutes Annotated §24-5-11.5-13 [Burns 1991 repl. Vol. and 1994 supp.])

Remodelers can assign the manufacturers' original purchase warranties directly to the owners. Assigning warranties transfers directly to the owners those warranties that the builders received when they purchased the goods for installation in the owners' homes.

Warning—Some manufacturers do not allow assignment of warranties on their products, and the assignment of warranties is governed by state law.

If remodelers intend to warrant the manufactured items or consumer products installed in the homes they remodel, the warranty must conform to the Magnuson-Moss Act guidelines (see Chapter 6 or this federal law).

The warranty limitations should also address environmental issues, such as indoor air quality, including radon, and possibly fiberglass (see Chapter 5 or these and other environmental liability issues). Remodelers must instruct all agents, representatives, and employees not to make any promises or representations to owners without specific authorization from the remodeler for whom they are working. In addition the contract should provide that the remodeler has made no guarantees, warranties, understandings, or representations (and that none have been made by any representatives of the remodeler) that are not set forth in the contract and the warranty. This section should be signed or initialed by the owner.

The remodeler's jobsite supervisors, managers, employees, and subcontractors who are likely to be in contact with the owners should sign agreements with their individual remodelers acknowledging that they are not authorized to make representations, promises, warranties, and the like without authorization of the remodelers.

Access to Home and Use of Utilities

Sometimes disagreements over minor matters such as access to the site and use of the telephone can ruin a congenial relationship between a remodeler and a client. To guard against potential misunderstandings and the resulting loss of goodwill and subsequent referrals, some remodelers include these items in their contracts.

Access—Owner shall grant free access to work areas for workers and vehicles. Driveways shall be kept clear and available for movement of vehicles during scheduled working hours, which will be ______ a.m. to ______ p.m. The remodeler and the remodeler's workers shall make reasonable efforts to protect driveways, lawns, shrubs, or other vegetation. However they shall not be responsible for damage to any of the items listed above unless the damage results from their gross negligence.

Inconvenience—The owner understands and accepts that during the course of the project inconveniences may occur from time to time, and the remodeler agrees to keep such inconveniences to a reasonable minimum.

Removal of Personal Property—The owner shall be solely responsible for removing or protecting personal property, inside and outside, and remodeler shall not be held responsible for damages or loss to personal property not removed from affected areas.

Removal of Material and Salvage—Material removed from the building during the course of the work shall be disposed of by the remodeler except those items designated in writing by the owner before work commences.

Clean-Up—The remodeler will leave the work site orderly at the end of each day. Upon completion of the project all remodeler's construction debris and equipment shall be removed by the remodeler and the premises left in neat, broom-clean condition, unless otherwise agreed upon herein.

Sign—The buyer agrees to permit the builder to display a sign on the site until the project is completed. [The parties might even agree upon the location of the sign.]

Telephone—The owner shall permit the remodeler and the remodeler's workers access to a telephone for local calls. No long distance calls will be made from the owner's home. [Some remodelers install their own telephones for large jobs or use a cellular phone.]

Toilet—The owner shall permit the remodeler and the remodeler's workers access to toilet facilities or compensate the remodeler for the cost of rented units.

Electricity and Water—Use of these utilities may need to be included in the contract as well. Otherwise these hidden expenses can be a jolt to the owner. For instance, prolonged use of heavy power tools or filling an in-ground pool that is part of a project can increase utility usage.

Photographs—Owner agrees to permit remodeler to take interior and exterior photographs of the residence for remodeler's use in marketing, for display, promotion, and advertising—

- at reasonable times
- upon giving the owner reasonable notice
- without compensation to the owner

Work Performed by Owner and Other Contractors

Some owners reserve the right to perform some of the work, or they reserve the right to subcontract part of the work. Some remodelers feel very strongly that the owner should not perform any of the work, and they discourage this practice, sometimes including such a warning in the contract.

If the owner reserves the right to perform some of the work or to award separate contracts in connection with other portions of the work, the contract must clearly describe the work to be performed by the owner or the subcontractor and the time within which the work is to be performed. Usually, if the owner enters into multiple contracts with different contractors for work on one project, the owner is under an obligation to coordinate and control the operations of all contractors to avoid unreasonable disruption of, or interference with, the operations of the other contractors. If that coordination fails the delayed performance may be compensable. Finally, if the owner performs certain work or has other contractors perform that work, the remodeler should expressly provide that he or she is not warranting such work.

Mandatory Clauses

Remodelers must take care to include certain mandatory language in their contracts. Several federal laws require that certain notices, warnings, or other information be included in sales contracts for custom and other new homes and in such contracts for consumer products, which often include remodeling contracts. State laws similarly often require that certain information be included in custom and other new home sales contracts and in such contracts for consumer products. For example, in at least one state, a home improvement contract must include the following statement next to the owner's signature line:

IMPORTANT NOTICE: You and your contractor are responsible for meeting the terms and conditions of this contract. If you sign this contract and you fail to meet the terms and conditions of this contract, you may lose your legal ownership rights in your home. KNOW YOUR RIGHTS AND DUTIES UNDER THE LAW.[38]

Warranties Under the Magnuson-Moss Warranty Act

If a remodeler gives a warranty and the warranty covers consumer products, the Magnuson-Moss Warranty Act applies. The Act does not require a remodeler to give a warranty. However it does regulate the form of whatever consumer product warranty a remodeler may voluntarily decide to give (see Chapter 6).[39]

Right of Cancellation and Rescission

Two federal consumer laws give homeowners 72 hours to cancel or rescind (terminate) certain transactions. Neither law is directed at remodeling projects in particular. However consumers entering into remodeling contracts may be covered by one or the other of these laws. Both laws require that consumers be advised in writing of the right to cancel or rescind a contract. Prudent remodelers—even if they are not sure that their contracts are covered by these laws—include the cancellation clause in their contracts or notify consumers of their right to rescind.

Right of Cancellation

Under this federal law, a consumer entering into a remodeling contract may be entitled to cancel the contract if (a) the remodeler personally solicits the sale and (b) the contract is made in the consumer's home or a location other than the remodeler's place of business. This law gives consumers a 3-day cancellation right. Compliance with this law requires including a notice of cancellation clause in the contract and following a few simple rules (Figure 3-2).[40]

Right of Rescission

This federal consumer protection provision applies to a credit transaction for which a security interest is or will be retained or acquired in a consumer's principal dwelling. The consumer may exercise the right to rescind until midnight of the third business day following consum-

Figure 3-2. Sample Notice of Cancellation

Notice of Right to Cancel Contract Secured by Your Home

Your Right to Cancel

You are entering into a transaction that will result in a lien and/or security interest on [or] in your home. You have a legal right under federal law to cancel this transaction, without cost, within 3 business days from whichever of the following events occurs later:

(1) the date of the transaction, which is ________; or
(2) the date you received this notice of your right to cancel.

If you cancel the transaction, the lien and/or security interest is also canceled. Within 20 calendar days after we receive your notice, we must take the steps necessary to reflect the fact that the lien and/or security interest on [or] in your home has been canceled, and we must return to you any money or property you have given to us or to anyone else in connection with this transaction.

You may keep any money or property we have given you until we have done the things mentioned above, but you must then offer to return the money or property. If [returning the property] . . . is impractical or unfair to you, [you] must offer its reasonable value. You may offer to return the property at your home or at the location of the property. Money must be returned to the address below. If we do not take possession of the money or property within 20 calendar days of your offer, you may keep it without further obligation.

If you cancel any property traded in, any payments made by you under the contract of sale and any negotiable instrument executed by you will be returned within 10 business days following receipt by the seller of your cancellation notice, and any security interest arising out of the transaction will be canceled.

How to Cancel

If you decide to cancel this transaction, you may do so by notifying us in writing at

____
(remodeler's name)

____
(street or Post Office business address)

____
(city, state, zip)

You may use any written statement that is signed and dated by you . . . [that] states your intention to cancel, or you may use this notice by dating and signing below. Keep one copy of this notice because it contains important information about your rights.

If you cancel by mail or telegram, you must send the notice no later than midnight of ____(date)____ (or midnight of the third business day following the later of the two events listed above). If you send or deliver your written notice to cancel some other way, it must be delivered to the . . . address [listed above] not later than that time.

I WISH TO CANCEL

______________________ ______________________
(consumer's signature) (date)

____

Source: *Contract for Repairs or Alterations* (Louisville, Ky.: Home Builders Association of Louisville, 1988), p.2

mation (the signing of the contract), delivery of the right of rescission notice, or delivery of the Truth in Lending disclosures, whichever comes last. The remodeler must deliver two copies of the notice of the right to rescind to each consumer entitled to rescind. Unlike the notice of cancellation, the notice of rescission must be on a separate document (Figure 3-3).[41]

Escalation Clause

If a party agrees to perform work for a fixed price, that party bears the risk of an increase in the cost of compliance because of an increase in the cost of labor or materials during the project. In other words a party will not be excused from performing the contract because it turns out to be more difficult or burdensome to perform than expected. The current saga of the increase in lumber prices is evidence that an increase in the cost of materials during the project can significantly increase the cost of compliance of a fixed-price contract using this alternate language:

> However the parties can provide in the contract that if a contract price increases because of any price increase in labor or materials or as a result of additional work required by the building department or other governmental agency before or during construction, the

Figure 3-3. Sample Notice of Rescission

Notice of Cancellation for Contract Solicited at Your Home Under Federal and Kentucky Law

Date____________________

You may cancel this transaction, without any penalty or obligation, within 3 business days from the . . . date [listed above]. If you cancel, any property traded in, any payments made by you under the contract or sale, and any negotiable instrument executed by you will be returned within 10 business days following receipt by the seller of your cancellation notice, and any security interest arising out of the transaction will be canceled.

If you cancel, you must make available to the seller at your residence, in substantially as good [a] condition as when it was received, any goods delivered to you under this contract or sale; or you may, if you wish, comply with the instructions of the seller regarding the return shipment of the goods at the seller's expense and risk.

If you do make the goods available to the seller and the seller does not pick them up within 20 days of the date of your notice of cancellation, you may retain or dispose of the goods without any further obligation. If you fail to make the goods available to the seller, or if you agree to return the goods to the seller and fail to do so, then you remain liable for performance of all obligations under the contract.

To cancel this transaction, mail or deliver a signed and dated copy of this cancellation notice or any other written notice, or send a telegram, to—

____________________ at ____________________
(remodeler's name) (street)

____________________ not later than midnight of
(city, state, zip)

____
(date)

I hereby cancel this transaction

____
(owner's signature)

Source: *Contract for Repairs or Alterations* (Louisville, Ky.: Home Builders Association of Louisville, 1988), p.2

owner will pay the remodeler the increase upon proof of such an increase. In addition the parties might agree that in the event of a price increase of more than a particular percentage, the owner has the option of paying the increase or terminating the contract after reimbursing the remodeler for all work completed before notice of termination.

Illustrative Case

The Einhorns entered into a contract with Ceran Corporation, a large residential builder, for the purchase of a townhouse to be constructed by Ceran. The contract, provided by Ceran, included a "price escalator" clause that gave Ceran the option of canceling the contract if the cost of labor or materials increased. The Einhorns had the option of paying any increased costs of labor and materials by providing notice within 10 calendar days after notice of any increase in cost. Subsequently Ceran notified the Einhorns it was exercising its right to increase the cost of the contract, but despite repeated requests by the Einhorns, Ceran refused to supply any information substantiating the need for the price increase. A closing took place by order of the court.

At that time the Einhorns paid the contract price to Ceran and paid the price increase to the clerk of the court, subject to a decision by the court. Finding for the Einhorns, the court held that the since the contract gave the Einhorns the option of paying the increase, Ceran had to furnish the buyers facts sufficient to enable them to determine whether the demand reflected real cost rises. The court observed that "Without some other verifiable or calculable standard, the buyer can be charged only the builder's actual cost increases and he must be prepared to demonstrate them sufficiently for the buyer to ascertain whether they are real."[42]

Uncovering and Correcting Work

Other topics typically included in general conditions include uncovering and correcting work, for example, work that (a) was covered before inspections occurred, (b) does not meet the contract standards, or (c) violates some other aspect of the contract. The general conditions can also provide for acceptance of work that was not done to contract specifications but that basically is acceptable to the client. The following sample language could also be adapted by the remodeler for use in the subcontractors' contracts to protect the remodeler.

> **Uncovering Work**—If work is covered in contradiction to the contract or the applicable laws, the remodeler shall uncover the work for observation or inspection and replace it without charge. If the homeowner or architect requests that work be covered that has been completed and inspected in compliance with the contract and local laws, the remodeler shall obtain a change order signed by the homeowner to charge the uncovering to the homeowner. If the work uncovered does not comply with the contract documents or applicable laws, the remodeler shall bear the costs of uncovering and redoing the work unless the homeowner caused the problem that needs to be corrected.
>
> **Correcting Work**—Upon written notice by the owner, the remodeler shall promptly redo and recover the work that does not meet contract specifications or applicable laws and pay the costs of redoing that work and recovering it, as well as any additional expenses for testing or inspections.
>
> If within a reasonable time the remodeler does not correct work that does not meet contract requirements or applicable laws, the owner may remove it and store (at the remodeler's expense) the materials that can be reused. If the remodeler does not pay the costs of removing and storing the materials within a specified period, the owner can sell the equipment and materials. If the proceeds of the sale do not cover what the remodeler would have paid to correct the work, the contract price shall be decreased accordingly. If the noncomplying work destroys or damages work done by the owner or another contractor, the remodeler shall bear the cost of correcting that work as well.
>
> **Acceptance of Nonconforming Work**—Instead of requiring removal of the work, the owner may choose to accept work that does not meet contract specifications so long as it conforms to the applicable laws. If the owner chooses to do so the total contract price will decrease accordingly, regardless of whether the owner has made the final payment.

Termination of the Contract

Occasionally unpredictable or uncontrollable events substantially change a construction project, and one party wants to be released from the contractual obligation. The remodeler and the owner can negotiate how to allocate responsibility for this possibility. Many contracts provide for termination for cause, such as the owner's nonpayment, the remodeler's substantial failure to comply with the contract terms, a substantial delay caused by a government entity that interrupts the work, a substantial cost increase resulting from the discovery of a hidden or unforeseen condition, or the like. In the absence of such a clause and unless the parties mutually agree to terminate, a party may not terminate the contract without liability, unless the other party defaults, and the default defeats the purpose of the contract. The sample language at the end of this section illustrates only one solution. The parties may agree to any of the following solutions if the owner terminates the contract without cause:

- The remodeler keeps any money earned up to the date of termination.
- The remodeler is entitled to a specified percentage of the remodeler's total anticipated profits.
- The owner must pay the remodeler for all work performed together with a sum of money equal to the amount of anticipated profits, or a reasonable profit on the entire agreement price.
- An owner must pay the remodeler a sum of money greater than the amount of anticipated profits. This provision should not call the additional fee a penalty because a court will invalidate a punitive damages clause in a contract. However a sum designed to compensate the remodeler for downtime while arranging for a new job may be justified.

The remodeler also may include a provision that allows the owner to cure a breach of contract within a specific time period by bringing all payments and other obligations up to date. This action permits the parties to reinstate the agreement so the remodeler can complete the job. Similarly, if the contract authorizes the owner to remove the remodeler from the job because of deficiencies in the work, the remodeler should include wording in the contract stating that the remodeler is entitled to notice of the deficiency and an opportunity to cure the deficiency before the termination is effective.

If the remodeler fails to substantially fulfill the contract terms, the termination provision may permit the owner to complete the work. In such a case the contract price would be reduced by the cost to the owner of completing the project. Additionally, if the cost of completing the project exceeds the contract price, the remodeler may be responsible for the additional amount. The parties also should allocate attorney's fees incurred in connection with termination of the contract. Another provision should address the event of bankruptcy of either party.

The contract should require that all termination notices be in writing (see Notices).

Remodeler's Remedies—If the owner (a) materially fails to comply with the provisions of this contract, (b) terminates the contract for any reason other than the remodeler's failure to perform, or (c) orders the remodeler or the remodeler's agents, employees, or subcontractors to stop work performed under this contract, the remodeler may—

- terminate this contract and retain any downpayments and deposits as liquidated damages
- recover all unpaid costs, expenses, and fees earned to the time of default or termination; the prorated cost of overhead expenses; and the costs, fees, and prorated overhead expenses for all change orders approved by the owner before termination
- institute judicial proceedings for specific performance and/or any other legal and equitable remedies.

(The costs specified in this section shall be based upon the costs specified in Section ____, Price, Deposit, and Payment, of this agreement.)

Owner's Remedies—If the remodeler fails to supply proper materials and skilled workers; fails to make payments for materials, labor, and subcontractors in accordance with their respective agreements; disregards ordinances, regulations, or orders of a public authority; or fails to materially comply with the provisions of the contract, the owner must give the remodeler written notice. After seven (7) days if the remodeler has failed to remedy the breach of contract, the owner must give a second written notice. If the remodeler still fails to cure the breach within seven (7) days after the second notice, the owner may terminate the contract. In such a case the remodeler shall reduce the contract price by the cost to the owner of finishing the work.

A material breach may include—

- persistent deviation from plans and specifications
- persistent disregard for applicable law
- repeated failure to maintain the schedule agreed upon

Arbitration, Mediation, and Other Alternative Dispute Resolution

During the course of construction, a disagreement between the parties to the contract may arise that they cannot personally resolve through negotiations. Because of the expense of a lawsuit and the long wait to have a case heard in court, litigation may not be a wise or efficient method of resolving the matter. Accordingly the contract should address how disputes will be settled. Arbitration and mediation are popular alternative dispute resolution (ADR) methods. For flexibility remodelers should consider including a clause in their contracts that allows them to choose arbitration, mediation, or litigation.

Arbitration

Binding arbitration is a process in which the parties submit their case to a neutral third person or a panel of individuals (arbitrators) for a final and binding resolution. Arbitration is a method for resolving disputes without the publicity of a lawsuit and generally costs less. Arbitrators usually are experts who may be more likely than a jury to understand the technical aspects of a construction controversy. Arbitration also often provides a speedier resolution than litigation. Additionally the mere existence of an arbitration provision may deter potential lawsuits.

States vary in their willingness to enforce arbitration provisions, but the clear trend is to uphold an arbitration agreement. Where materials for a house come across state lines, the Federal Arbitration Act (FAA) may govern. Both the FAA and the individual state acts provide that if the contract has an arbitration clause, one party may force the other to arbitrate. Remodelers should suggest arbitration or mediation procedures early in a dispute, before an owner begins to suffer mental anguish that a jury might find worthy of compensation with a damage award.

In a minority of jurisdictions, an arbitrated case can still be brought to court, even if a contractual provision states that the arbitrator's decision is binding. In the majority of states, however, the arbitrators' decision is final and binding, and neither party can appeal it except in the case of fraud on the part of the arbitrator. Although the parties in these jurisdictions may attempt arbitration before filing a lawsuit, if one party decides to circumvent the process, a court may allow it.

A formal dispute resolution organization, such as the American Arbitration Association (AAA) can conduct an arbitration proceeding. The AAA uses its *Construction Industry Arbitration Rules.*

An AAA proceeding under the *Construction Industry Arbitration Rules* can be expensive because the parties pay both the AAA and the arbitrator(s) of the dispute (generally the AAA assigns three arbitrators for disputes over $100,000 and only one for cases under $100,000).

Alternatively the parties can provide tailor-made arbitration rules either in the contract or in a separate contract document and thus eliminate the need for an arbitration association. For example, the parties can agree that each party to the contract independently will select an arbitrator, and those arbitrators jointly will choose a third.

The contract also may—

- provide for discovery and application of the rules of evidence (During formal discovery each side unearths facts and documents from the other side that may be helpful in defending or prosecuting its case. Discovery is often expensive and time-consuming. It includes depositions, interrogatories—written questions and answers—and production of documents.)
- set time limits for presentation of each party's case
- limit the maximum damages allowed in the arbitrator's decision and award

Rather than develop their own dispute resolution procedures, the parties can agree to follow the procedures outlined in the federal Magnuson-Moss Warranty Act, 15 United States Code sec. 2301 et seq. (see also 16 Code of Federal Regulations sec. 107).

Remodelers should not mention alternative dispute resolution in their warranty provisions because they may be forced to use the Magnuson-Moss settlement procedures.

The AAA can be contacted at the address listed below:

American Arbitration Association
140 West 51st Street
New York, NY 10020
(212) 484-4000

All disputes between the parties to this contract arising out of or related to any contract term(s), or any breach or alleged breach of this contract, will be decided by arbitration unless the parties mutually agree otherwise in writing.

The arbitration shall be conducted by (specify the organization or named arbitrator[s] agreed upon) in accordance with [the rules adopted by the arbitration body chosen, the rules specified below, or the alternative dispute resolution proceeding specified in the federal Magnuson-Moss Warranty Act, 15 U.S.C. §2301 et seq. (see also 16 C.F.R. §107)]. The parties reference the Magnuson-Moss Act only to provide rules of arbitration in the event of a dispute, and they specifically do not incorporate the Magnuson-Moss warranties into this contract.

The parties must file a written notice of arbitration with the other party to this contract and with (the arbitration association or arbitrator[s] chosen. The notice of arbitration may not be filed after the date that a claim based on the dispute would have been barred in a judicial proceeding by the applicable statute of limitations or repose (cessation of activity).

Either party may specifically enforce (a) a decision made under this agreement to arbitrate or (b) any valid agreement to arbitrate with additional persons, under applicable arbitration laws. The decision made by the arbitrator(s) will be final and binding, and any court with jurisdiction over the decision may enter a judgment upon the arbitrator's decision.

Mediation

Mediation, like arbitration, is a process whereby the conflicting parties meet voluntarily to negotiate a private and mutually satisfactory agreement aided by a neutral third party. A key difference between the two methods is that unlike an arbitrator a mediator does not make a decision in favor of one party or the other. Instead a mediator focuses on negotiation and problem solving and assists the parties in this process. Remodelers' mediation may follow the *Construction Industry Mediation Rules* of the American Arbitration Association. If mediation has a down side, it may be that if mediation is not successful the parties are back where they started, and they may have to resort to litigation or another form of alternative dispute resolution. For this reason the parties may want to include a provision in the contract making all communications during the mediation confidential.

> If a dispute arises between the parties relating to or arising out of any provision of this agreement or any breach or alleged breach of this contract, either party may request mediation. Either party may invoke the dispute resolution procedure of this clause by giving written notice to the other. The notice should include a brief description of the disagreement. A mediator will be selected by mutual consent of the parties. The parties agree to participate in good faith in the mediation to its conclusion as determined by the mediator. No party will be obligated to continue in the mediation if a solution has not been reached and put in writing within [number of days] of the first mediation session. The costs of mediation, including fees and expenses, shall be borne equally by the parties.

At this point the clause would include the sample language from the arbitration clause printed above beginning with the phrase, "The arbitration shall be conducted by . . ." and ending with the phrase "and any court with jurisdiction over the decision may enter a judgment upon the arbitrator's decision."

Attorney's Fees

In the United States each party to a lawsuit generally pays its own attorney's fees—win or lose. Thus the prevailing party in a lawsuit often ends up paying a considerable amount of the judgment to an attorney, or pays a considerable sum of his or her own money to the attorney for successfully defending a suit. One way the parties can avoid such an expense is to contract for the award of attorney's fees to the prevailing party in any action arising out of the project. If the contract includes such a clause the decision to pursue a case may be made strictly on the merits of the case, whereas if the contract is silent on this matter the party must factor in the cost of the attorney's fees in deciding whether to pursue the claim. Moreover the inclusion of such a clause may discourage frivolous suits and may force the parties to deal more forthrightly with each other. On the other hand, some attorneys do not favor including such a clause in the contract because they fear that it will have the opposite effect and that it will encourage litigation.

> If either party to needs to enforce provisions of this contract or to obtain redress for the violation of any provision hereof, whether by litigation, arbitration, or otherwise, the prevailing party shall be entitled to any reasonable attorney's fees, court costs, or other legal fees incurred herein in addition to any other recovery obtained in such action.

Entire Agreement and Severability

This provision emphasizes that this contract supersedes all previous agreements between the parties. It also assures that if a court determines that one provision is unenforceable, the remainder of the contract will remain in effect.

Finally the parties can agree not to assign or sell their rights and responsibilities in the contract to another party. The assignment clause restricts the owner's right to sell the contract (and its payment obligations) to another person who may not have adequate financial resources to pay the remodeler.

> This contract (including the documents incorporated by reference) constitutes the entire agreement between the parties. It supersedes all previous or current agreements and understandings whether written or oral. This contract may be changed only by a written contract amendment signed by all parties to this contract. The contract shall be binding upon each of the parties' respective heirs, executors, administrators, successors, and assigns. However this contract shall not be assigned without the written consent of all parties.
>
> Each provision of this contract is separable from every other provision of the contract, and if any provision is unenforceable or revised, the remainder of the contract will remain valid and enforceable.
>
> This contract will be governed by the law of the jurisdiction in which the property is located.
>
> The provisions of this contract shall survive the execution and delivery of the deed and shall not be merged therein.

Specifications

The general specifications are usually prepared as a separate document and referenced in the contract along with other documents, such as plans and architectural drawings. The specifications should state that they cover labor and materials for construction of an addition or alteration to a specific building, and they should include a legal description of the property as it appears in the contract (see Section ____, Caption). Generally the specifications would include the following statement: "Unless otherwise specified in the contract, the specifications are as listed" (Figure 3-4). The specifications should be as detailed as possible and reference plans and blueprints where appropriate. Figure 3-4 is provided only as an example. Other necessary parts of the construction process may not be listed here, and some of the listed items may not apply to a particular project. All work must be listed and described in the specifications document.

Description of Work to Be Performed

This section should describe the project in detail, as set out in the other contract documents. It should include a specific description of what a remodeler will do and not do. This last point is particularly important with regard to those items that were discussed and priced but that were ultimately rejected.

The cost of the work and the remodeler's fee are based on the work described here. This section describes the scope of work in sufficient detail to clarify what will constitute a change by change order. For example, the remodeler's scope of work may include compliance with the local building code or other laws and regulations. The description of work to be performed—

- specifies what the remodeler will do and what he or she will not do
- may describe the portions of the project for which each of the other parties is responsible
- spells out how the work is to be performed and describes the remodeler's responsibility for subcontractors
- specifies the kind or brand of materials to be used
- specifically identifies any unusual responsibilities that the owner may request of the remodeler

The remodeler's scope of work shall not include the following:

- detection, treatment, encapsulation, or removal of asbestos or other hazardous materials (for a more complete discussion of asbestos, see Chapter 5)

Figure 3-4. Items Usually Included in Specifications

- Permits and inspections
- Excavating
- Footings and foundation
- Termite treatment
- Rough framing
- Roofing
- Gutters and downspouts
- Windows
- Doors
- Plumbing
- Electrical
- Heating, ventilation, and air-conditioning
- Drywall
- Ceramic tile
- Closet finish
- Interior trim
- Paneling
- Ceilings
- Painting and staining
- Finish floors
- Finish hardware
- Landscaping
- Venting
- Carpet
- Walks and driveways
- Bath fixtures
- Kitchen cabinets
- Countertops
- Kitchen appliances

- correction of damage caused by termites, dry rot, or similar cause
- changes to existing wiring in areas undisturbed by alterations

Illustrative Case

In this case Spotswood I. Quinby, Inc., promised to construct and sell to Denice a house with a finished recreation room in the basement in a fashionable section of Montgomery County, Maryland, While visiting the premises, Denice inspected the work in the recreation room and noticed that the ceiling was low. Later through his attorney, Denice wrote the builder to complain that the height of the recreation room was only 6 feet 9 inches and that the BOCA Code (which by ordinance was the Montgomery County building code) required not less than 7½ feet.

The contract was silent as to the height of the recreation room. The parties were unable to resolve the problem, and Denice sued the builder, alleging in part that the finished recreation room did not conform to the county building code.

Finding for Denice, the court relied on the general rule that unless a contract provides otherwise the applicable law is read into and becomes a part of the contract. The court concluded that compliance with the building code was an implied condition of the contract and that the builder's failure to comply with the code justified Denice's refusal to complete the deal. The court noted, however, that only a substantial noncompliance with the provisions of the building code would excuse performance.[43]

Conformance with Plans and Specifications; Changes in Work, Property Lines, or Matching Materials

Remodelers must explain to the owners that the actual work may deviate somewhat from the plans and the specifications. For example, remodelers should explain to owners that they cannot guarantee that new materials, textures, colors, and planes will exactly duplicate the existing ones. Similarly remodelers may want to protect themselves in the event of deviations from property lines, survey lines, and the like. These facts should be disclosed to the owner in conversations at the beginning of the marketing and negotiation processes. In addition these provisions should be included in the contract and in the certificate of acceptance to be signed or initialed separately by the owner as evidence that the remodeler disclosed the possibility of deviations or inconsistencies to the owner. The owner should initial each of the provisions in the sample language printed immediately below.

Remodeling Work—The remodeling work shall be constructed to conform substantially with the plans and specifications that are set forth in the attached Exhibit(s) (include numbers and names of exhibits), the terms of which are incorporated in this contract by reference). The owner acknowledges that during the course of the work certain changes, deviations, or omissions may be necessary because of (a) the requirements of governmental authorities having jurisdiction of the property, (b) suggestions and design changes made by the architect, or (c) particular conditions of the job. The owner hereby authorizes the remodeler to undertake, without the need for specific authorization, any changes, deviations, or omissions required by governmental authorities, or authorized by the architect. Changes in the particular conditions of the job shall be handled in accordance with Section ____, Changes in Work and Change Orders, of this contract.

Property Lines—Owner shall locate and point out the property lines to the remodeler. When necessary, owner shall furnish, at owner's expense, all necessary surveys. Owner assumes all responsibility for the accuracy of all boundary markers.[44]

Matching Materials—The remodeler calls attention of the owner to the limitations of plaster and stucco. While the remodeler shall make every effort to blend existing textures, colors, and planes, exact duplication is not guaranteed. Roofing materials and/or the color of the roofing material may not match existing roofing because of manufacturer's discontinuance of color or product, shading difference of product, age, or changes in the product or the product's manufacturing process.[45]

Four • Design-Build Contracts Used by Remodelers and Custom Builders

Design-build describes a project in which the custom builder or remodeler designs a job specifically for a particular client and constructs the job based on that design. The design and construction functions are undertaken by one entity, and this arrangement distinguishes the process from the traditional method of construction, in which the owner contracts directly with the designer for the design of the project and directly with the builder or remodeler for construction of the project. Discussing the pros and cons of the design-build arrangement is beyond the scope of this book.[46] However this chapter shows custom builders and remodelers how to avoid two potential pitfalls that await the uninitiated design-build company: practicing architecture illegally and offering design services without adequate liability protection.

Practicing Architecture

Custom builders or remodelers who offer design services to their customers should verify that they have the proper license to do so, or that the work is being performed by a licensed architect or designer. All states have regulations governing the practice of architecture and almost all prohibit unlicensed persons from offering architectural services.

However the states differ as to what services constitute the practice of architecture. The majority of states hold that design services, such as the preparation of schematics, working drawings, plans, and specifications are architectural services and may be performed only by licensed architects. Builders may prepare preliminary sketches for the architect or designer to use as the basis for the schematics, drawings, and floor plans.

Practicing architecture without a license may be costly; in many courts a contract performed by a person unlicensed to perform such services is unenforceable. Figure 4-1 lists licensing issues that design-builders and design-build remodelers should discuss with their attorneys.

Exemptions

Fortunately, most states have exemptions to their licensing or registration requirements, and these exemptions differ widely. Structures with certain uses (often single-family homes), sizes, and costs may be exempted. Builders and remodelers must look to their own state's architectural laws to see what exemptions, if any, the state permits.

Stamping or Sealing Plans

The architectural licensing laws cannot be circumvented by having an architect seal or stamp plans prepared by a design-builder or design-build remodeler. This practice is illegal in the vast majority of states. A builder or remodeler who obtains such a seal may be guilty of a misdemeanor. Some states consider a violation of an architectural regulation a civil offense; others view it as a criminal offense, punishable by a monetary fine and/or a short jail term. Thus the prudent builder or remodeler will consult an attorney experienced in construction law before offering design services. An architect or building designer might provide the services as a member of a custom builder's or remodeler's staff, as an independent supplier of design services under subcontract to the builder or remodeler, or as part of an alliance in which the architect or designer and the builder or remodeler feed each other work. However in some states an engineering seal applied under a similar arrangement is sufficient.

When the designer is a subcontractor to, or a joint venturer with, the builder or remodeler, the parties should have a written contract outlining their respective responsibilities and liabilities. For example, in addition

Figure 4-1. Licensing Issues Design-Builders and Design-Build Remodelers Should Discuss with Their Attorneys

- Is design-build specifically addressed by statute?
- Must a design-build joint venture get a special license, or will the individual licenses of the designer and the builder or remodeler meet the statutory requirements?
- Do the licensing statutes for architects, engineers, and contractors indicate how the design-build entity needs to be structured in order to comply with the statutory licensing requirements?
- Will a licensed architect on staff meet the licensing requirements for design by a design-build firm?
- Can a homeowner contract for design-build services directly with a licensed designer who does not have a contractor's license or directly with a licensed contractor who does not have an architect's license?
- Will the requirements for a licensed architect be satisfied by subcontracting the architecture to a licensed architectural firm if the design-builder or design-build remodeler is not licensed as an architect or as an engineer and does not have an employee who is licensed as such.

Source: Reprinted with permission from Thomas H. Asselin and L. Bruce Stout, *Legal Exposure of the Design/Build Participants: The View of the General Contractor,* The Construction Lawyer, Vol. 15, No. 3 (August 1995), p. 8–28.

to the standard contract provisions discussed in Chapter 8 (cost of the work, the payment terms, time for performance, and so on), the contract might include clauses making the design subcontractor solely responsible for particular aspects of design, including stamping relevant contract drawings, drafting the accompanying specifications, and indemnifying the contractor for design errors of the design subcontractor.[47]

Protection Against Liability for Design Errors

The design-builder or design-build remodeler may not be adequately insured. Design errors may or may not be covered under a construction policy (for example, a general liability policy) unless the policy expressly states that it covers professional liability, which is unlikely. If the policy does not contain an exclusion for professional liability, the insurance company might pay. Or a court might rule in the custom builder's or remodeler's favor because the policy does not exclude coverage for design services. Neither result is certain, however, and the wise custom builder or remodeler would not wait until a problem occurs to find out.

To fill this gap in insurance, design-builders and remodelers should consider obtaining an errors and omissions policy. However these policies are extremely difficult to obtain and prohibitively expensive. Such insurance is far more expensive for builders than it is for architects and designers, who are designing every day. Thus the builder or remodeler who is working with an unusual structural design or who has questions about the design needs to have the structural design reviewed by a structural expert.

Alternatively builders and remodelers should consider allocating this risk to a third party. For example, a builder or remodeler might enter into a consulting agreement with an architect or an engineer to review the plans and specifications. The consultant could agree to indemnify the design-builder or design-build remodeler for damages their consultations might cause, but this solution is not realistic. The builder or remodeler should obtain the consultant's malpractice insurance certificate or a copy of the insurance policy. The design-builder or design-build remodeler could also ask to be listed as an insured on the consultant's insurance policy.

Design-Build Contracts

A design-builder or design-build remodeler may need three contracts to protect his or her interests adequately: a contract for a feasibility study, a design contract, and a construction contract. Sometimes the feasibility and design phases are included in one contract. Other times the design and construction phases are combined. This section focuses on the design contract. For tips on drafting a design-build construction agreement consult Chapters 2 and 3.

Design Contract

The design contract is sometimes called a preliminary design build-agreement or a preconstruction services contract. It covers the scope of the design work to be performed, the cost of the design service, and the rights and duties of the parties. This contract serves the following purposes:

- Sets forth the parties to the contract, their addresses, phone numbers, and so on.
- Briefly describes the project.
- Specifies the scope of the design. These specifications may take the form of a phase schedule. For example—

  Phase 1—Schematics
  Phase 2—Preliminary design and floor plan
  Phase 3—Working drawings and specifications

- Specifies the cost of design services, including the amount, manner, and time of payment. If the work is to be done in phases, payment may be linked to the start of a new phase. It may also be tied to levels of service. For instance:

  Level 1—Thumbnail sketches and one visit to the site might be free.

  Level 2—Any out-of-pocket expenses for engineering or other fees would be paid by the client.

  Level 3—Includes everything up to and including working drawings for a specific fee or hourly rate plus out-of-pocket expenses.

  In another example, the initial schematics for a $2,000 design job might cost $700. The payment schedule, however, might call for payment of $400 for the schematics and divide the remaining $300 between phases 2 and 3. The low initial design fee gives the design-builder or design-build remodeler time to let the design itself entice the customer into deciding on the whole job:

Phase 1—Schematics	$400
Phase 2—Preliminary design and floor plan	$800
Phase 3—Working drawings and specifications	$800

- Explains how outside services, such as engineering, will be handled and charged to the client.
- Provides a method of payment in the event the contract is terminated during the various phases of preconstruction.
- May include a liquidated damages clause.
- Explains what will happen if the client materially changes the scope of the work or the manner of its execution.
- Includes a completion date for the design phase. It may make the date contingent on the owner (a) providing necessary information and (b) making necessary decisions within a reasonable time.
- Specifies whether the design fee will be credited against the construction contract price in the event the client uses the firm for construction.
- Explains who owns the drawings, specifications, and other documents and how they can be used. For example the contract might specify that the plans remain the sole property of the builder or remodeler and that the purchaser may use them only for construction, repair, alteration, or other improvement by the builder or remodeler.
- Discusses the client's dual rights of rescission and cancellation. (A sample design contract appears in Figure 4-2.)

Figure 4-2. Sample Design Contract

This preconstruction agreement is made this ___________ day of ________, 19___, between __________________, the client, who resides at ______________________, and _____________, the [builder or remodeler], whose principal place of business is at __________. The client and the [builder or remodeler] agree as set forth below:

Project: (describe in detail) This project is designed to fall within a budget range of $_____ to $_____ for construction.

1. **Furnishing Design**—The [builder or remodeler] shall furnish the design of the project in accordance with the following schedule:
   Phase 1—Based upon the client's project requirements, the [builder or remodeler] will provide design schematics, including field measurements. These schematics are to assist the client in determining the feasibility of the project.
   Phase 2—Upon approval of the design schematics, the [builder or remodeler] will proceed with preliminary design development, floor plan, and elevations. Unless otherwise noted, drawings will be to scale.
   Phase 3—From approved preliminary design documents the [builder or remodeler] will provide working drawings and specifications. These working drawings will serve as the basis for the [builder's or remodeler's] estimate of the cost of construction and for the construction of the project.
2. **Payment Schedule**—In return for the foregoing services, the client agrees to pay the [builder or remodeler] as follows:
   Phase 1—Compensation for phase 1 shall be in the amount of dollars ($____), and the payment is due on ___________.
   Phase 2—Compensation for phase 2 shall be in the amount of dollars ($____), and the payment is due on __________.
   Phase 3—Compensation for phase 3 shall be in the amount of dollars ($____), and the payment is due on __________.
3. **Engineering and Other Extra Fees**—The [builder or remodeler] anticipates that the scope of the project will require the services of an engineer. The client shall pay the engineering fees at ________ dollars ($____) per hour, up to a maximum cost of _________ ($____). The [builder or the remodeler] shall not be responsible for the payment of any engineering fees.
4. **Failure to Proceed**—If the client elects not to proceed after any phase, the [builder or remodeler] is entitled to payment in accordance with paragraph B.
5. **Change in Work**—If the scope of the work or the manner of its execution is materially changed, the additional work shall be billed on an hourly basis at the following rates:
   [builder or remodeler] @ _________ dollars ($____); engineer @ _________ dollars ($____); others [list them] @ _________ dollars ($____).
6. **Ownership of the Documents**—The preliminary design documents, working drawings, and specifications are for the sole use of the [builder or remodeler] in connection with this project, and they shall remain the property of the [builder or remodeler]. They are not to be used by the client or any third party on other projects without the written consent of the [builder or remodeler]. The client will be liable to the [builder or remodeler] for all losses arising out of the unauthorized use or sale of these copyrighted documents.
7. **Client Cooperation**—Client will provide full information regarding the owner's requirements for the project and make necessary decisions required for completion of design documents in a timely manner.
8. **Delivery of Estimated Costs**—The [builder or remodeler] shall complete the design phase and provide the client with the cost to construct the project by the _________ day of ____________, 19____.
9. **Client's Right to Cancel**—Pursuant to federal law you have three (3) days from [signature date] to cancel this contract. See the attached form for an explanation of this law.
10. **Client's Right of Rescission**—Pursuant to federal law you have three (3) days from [signature date] to rescind this agreement. Client acknowledges receipt of two (2) copies of the notice of the right of rescission.

____
(buyer's or owner's signature)

____
(name of builder, remodeler, or corporate name where applicable)

____
(buyer's signature)

By ______________________
(authorized signature)

Date ______________________

Title ______________________

Date ______________________

This contract is dated, and becomes effective,

____
(month, day, year)

____
(buyer's initials)

Five • Environmental Liability

Speculative builders, custom builders, and remodelers are exposed to several sources of potential liability for environmental hazards to the public or for health risks to the buyer. This chapter discusses three issues that builders and remodelers can address in their contracts: contaminated purchased property, the owner's contaminated property, and indoor pollution.

The hazardous waste laws of the state and federal governments cover a wide range of chemicals and products. A builder's or remodeler's potential liability to these governments for hazardous waste clean-up is significant enough to warrant discussion here. Builders and remodelers should recognize that in addition to hazardous waste liability to the governments (which are discussed later in this chapter) consumers may file a lawsuit against a builder or remodeler if a home is found to be contaminated by hazardous waste or toxic materials.

However such cases are based on particular state tort laws (for example, the laws of negligence, strict liability, and related legal theories). Therefore they are beyond the scope of this book. Other major environmental issues that present potential liability for builders, developers, and remodelers, but that are beyond the scope of this book, include compliance with the following:

- federal, state, and local laws concerning the preservation of wetlands or other environmentally critical areas
- occupational, safety, and health regulations that are designed to protect employees and the community at large
- a host of other laws regarding clean air, clean water, coastal zone management, and the handling and storage of toxic materials

Failure to comply with the applicable laws can bring upon a builder, developer, or remodeler severe penalties and damages. Indoor pollution, a major environmental issue of concern to builders and remodelers, includes the potential liability to consumers for exposure to radon, asbestos, lead, and other indoor air pollutants. Though indoor air contains numerous potential pollutants, radon has taken a front seat in the media and before environmental and health safety agencies. Because radon is the issue that consumers inquire about most often, this chapter includes a discussion of radon, how it affects (and should not affect) the real estate transaction, and how builders should treat the potential liability stemming from radon. This chapter also discusses liability problems associated with two other indoor air pollutants, asbestos and lead; their importance to remodelers is also discussed here.

Hazardous Waste Liability

Builders and Developers Buying Property

This section explains potential liability for hazardous waste contamination on property builders and developers own (even if only temporarily). The sample contract language presented here may be included in land purchase agreements to limit potential liability.

The Comprehensive Environmental Response, Compensation, and Liability Act (CERCLA)—also commonly referred to as Superfund—authorizes the federal government to require four categories of potentially responsible parties (Figure 5-1) to pay for the release or threatened release of hazardous substances into the environment.

Any one of the parties in Figure 5-1 may be held liable for the entire clean-up cost regardless of negligence (or lack of negligence), amount of contribution to the contamination (if any), and compliance with laws in effect at the time of disposal.

CERCLA provides an "innocent purchaser" defense for property owners who can prove that they did not know and had no reason to know that any hazardous substance contaminated the property. To qualify for the defense, the party "must have undertaken, at the time of the acquisition, all appropriate

Figure 5-1. Potentially Responsible Parties

- The current owner or operator of the contaminated site
- Any person who operated a facility or owned a site when toxins were deposited
- Any person who created and disposed of the hazardous substances
- Any person who selected the site and deposited, treated, or transported the waste

inquiry into the previous ownership and uses . . . in an effort to minimize liability."[48]

CERCLA also provides a defense for contamination resulting solely from an act of God, an act of war, and acts or omissions by third parties not in a contractual relationship with the landowner. Persons who acquire property through inheritance or bequest are similarly exempt from liability.

Many states have enacted Superfund-like statutes that impose liability on landowners for the clean-up of hazardous waste. The most burdensome regulatory and liability schemes require sellers to remove contamination before the property is transferred and to notify purchasers or the state before transfer.

Some states also have enacted "superliens" that enable the state to place a lien, superior to all others, to secure payment for the clean-up. In a few states, liens can be placed on all of an owner's real property in the state—whether or not it is polluted—to pay for the cost of cleaning up a contaminated parcel.

Hazardous waste liability is not limited to properties bordering industrial sites. Rural and suburban properties and groundwater are often contaminated by pesticides, seepage from underground storage tanks, and improper disposal of industrial or drycleaning chemicals. Depending on the size of the lot or project, the prudent builder or developer may want to hire a consultant to conduct an environmental assessment or audit of the site. The steps listed in Figure 5-2 will help to minimize the risk of purchasing contaminated property and to limit a builder's liability in the event the property is contaminated.

Illustrative Case

Subdivision property owners brought an action against a lending institution, residential developers, construction companies, and real estate agents and agencies that participated in the development of a subdivision on a site formerly used as a wood-treatment facility. The property owners complained that highly toxic waste accumulated on the property. They invoked CERCLA and the Resource Conservation and Recovery Act (RCRA) in seeking damages and response and cleaning costs.

Although the defendants had not introduced the toxic waste at the site, certain of them had filled in and graded pools containing hazardous waste. One issue before the court was whether the defendants owned or operated the property at the time of disposal of the toxins. The court held that "disposal" meant the "discharge, deposit, . . . or placing of any solid waste or hazardous waste . . . into or on any land . . . ," and that disposals may result when hazardous materials are moved during landfill excavations and fillings. Another issue before the court was whether filling and grading the pools constituted "treatment" of hazardous waste subject to CERCLA regulation. The court noted that the term "treatment" included any activity designed to change the physical form of hazardous waste. Thus the court concluded that the defendants who filled and graded the pools might be shown to be arrangers for, or transporters of, toxic materials.[49]

A real estate purchase agreement should allow sufficient time for the buyer to inspect the land for hazardous waste through an environmental audit. The audit generally reviews past practices, business records, and government documents. If warranted, actual physical inspection, testing, and sampling also ought to be performed.

The buyer can require the seller's cooperation with the investigation. The standards and procedures of the audit should be set out specifically in the sales contract. The agreement should also give the buyer the option of terminating the contract upon discovery of a hazardous waste problem.

Title insurance generally does not protect real estate purchasers against hazardous waste liability. Buyers should review records in both the U.S. District Court and the state court in the jurisdiction in which the property is located to find out if the property is burdened

Figure 5-2. Steps for Recognizing a Contaminated Site

- Inspect the property and surrounding area.
- Note any discolored or defoliated soils or vegetation, odors, chemical drums and other containers, corroded drains, stained pavement, or discolored surface water.
- Look for storage tanks; industrial, mechanical, and electrical equipment; manhole covers; loading docks; fill pipes; and vent pipes.
- Investigate the historical use of the property:
  - —Has the property been the site of commercial, industrial, agricultural, or residential use?
  - —Identify potential hazardous or toxic materials associated with the past or present use of the land.
  - —If the land is vacant, determine whether it has ever been used as a landfill or storage area.
- Are hazardous waste problems common to the area?
- Review any easements on the property to determine whether easement holders may have used hazardous or toxic materials. Look for railroad easements, pipeline easements, or mineral rights granted to mining or oil companies.
- Determine and evaluate all occupants of the property. Do not limit the inquiry to previous record owners of the property.
- Consult with local, state, and federal environmental agencies to determine if the site is on a list of properties suspected of contamination or awaiting clean-up.
- Check for the following government permits issued to current or previous owners that could indicate the use of toxic chemicals:
  - —underground storage tank permits
  - —above-ground storage tank permits
  - —hazardous waste permits
  - —solid waste facility permits
  - —well-abandonment permits
  - —well-appropriation permits
- Review records in both the U.S. District Court and state courts for the jurisdiction in which the property is located to find out if the property is encumbered by any existing state or federal liens.

with any existing state or federal hazardous waste liens because these liens may not be recorded in local (county or state) recorder of deed offices.

Finally the buyer should seek indemnification provisions that will hold the seller liable for any costs related to unforeseen hazardous waste problems. The government can and will sue any of the four categories of responsible parties for clean-up costs. However an indemnification clause can establish the right of the buyer to recover liability costs eventually from the seller. An indemnification clause should include out-of-pocket costs and should not place a dollar cap or time limit on the indemnification. Builders should pay careful attention to the seller's financial ability to back the provisions. If the parties agree on a specific monetary amount for which the seller will be responsible, the buyer may consider requiring the seller to post a bond for the duration of the expected clean-up.

Inspection Provisions and Option to Terminate Contract

Buyers should incorporate into their land purchase agreements a provision that allows additional time, if needed, to conduct an environmental audit. A buyer may decide, after physical inspection of the property or a review of the chain of title, that toxins may have been disposed of on the site and that a study should be conducted to verify or discount this suspicion. Three options are provided in the sample language below. The language in brackets [ ] shows the reader ways the provision could be modified.

Option 1—[At any time prior to the closing of title or for ______ ( ) days following the execution of this contract] the buyer may enter upon the premises for the purposes of investigating the property or conducting an environmental audit to determine whether hazardous waste or toxic substances are present on the site. The terms *hazardous waste* or *toxic substances* in this agreement have the same meaning that they have in these state and federal laws: [List the laws.]. If the inspection discloses the presence of hazardous waste or toxic substances and the seller is unwilling to correct the problem to the satisfaction of the buyer, the buyer or seller may cancel this contract and all rights of both parties under the contract shall cease. The canceling party must serve a written notice of cancellation upon the other party or the other party's attorneys either in person or by registered mail.

The seller shall return to the buyer ________ percent (____%) of the deposit [with interest at ____ percent (%) per annum)]. Costs and damages shall be allocated as follows: [List costs and damages and how they are to be allocated.]

Option 2—The seller hereby grants to the buyer and the buyer's agents and employees the right to enter the property or any portion thereof [at any time prior to the closing of the title or for ____ ( ) days following the execution of this contract] for the purpose of conducting at the buyer's own cost any soil, geologic, engineering, or environmental investigations that the buyer may desire.

If any of these investigations disclose the presence of hazardous waste or toxic substances and the seller is unwilling to correct the problem to the satisfaction of the buyer, the buyer or the seller may cancel this contract and all rights of both parties under the contract shall cease. The canceling party must serve a written notice of cancellation upon the other party or the other party's attorney either in person or by registered mail.

The buyer agrees to indemnify and hold the seller harmless from any costs or liability incurred by any investigations conducted pursuant to this provision. If this contract is voided for any reason not caused by the seller, the buyer, at the written request of the seller, shall repair any damages caused to the property by these investigations and restore the property to the condition it was in before the investigation began.

Option 3—The buyer shall promptly cause (designate testing facility), the testing facility, to inspect and, if necessary, test the soil, subsoil, water, and air conditions of the property to ascertain whether the property contains hazardous waste or toxic substances as defined under state or federal law. The testing facility shall determine the number and quality of tests required to ascertain the presence or absence of hazardous waste or toxic substances. The buyer agrees to furnish a copy of the report of the testing facility to the seller on or before (date). If the report indicates that, in the judgment of the testing facility, the property contains hazardous waste or toxic substances, the buyer shall have the option to terminate this agreement. The buyer shall deliver written notice of the option to terminate to the seller within ____________ (____) days of receipt of the report of the testing facility. Costs and damages will be allocated as follows: [List costs and damages and how they will be allocated.].

Cooperation from Seller

The buyer may want to include in the contract a provision obligating the seller to cooperate in an investigation of the property.

> The seller agrees to sign and execute any and all documents that may be required by any person, firm, partnership, company, or local, state, or federal agency as a part of conducting an environmental audit of the property to determine the presence or absence of hazardous waste or toxic substances.

Seller Warranties

If the seller warrants that no hazardous substances are on the property, but they are indeed present and the buyer becomes liable for clean-up costs, the buyer can recover those costs from the seller. The contract should specify that the warranty in the contract will survive the closing.

> The seller warrants that the use, maintenance, operation, or condition of the property complies with all regulations, statutes, rules, and codes of all local, state, and federal governmental agencies having jurisdiction over the property.
>
> The seller further warrants that the property does not contain any hazardous waste or toxic substances that may create liability for the purchaser or its agents, heirs, or assigns under local, state, or federal laws.
>
> The provisions of this contract of sale will survive the real estate closing and shall not be merged with the deed.

Indemnification from Seller

The buyer also may be able to negotiate a specific provision in the contract that allocates responsibility to the seller for any liability imposed on the buyer for hazardous waste clean-up costs. As a practical matter the effectiveness of such a provision depends on the seller's ability to pay for such a clean-up.

> The seller agrees to indemnify and hold the buyer harmless from any costs or liability incurred by the buyer or the buyer's agents, heirs, or assigns for damages paid to any local, state, or federal agency or other government or private entities for costs of cleaning up hazardous waste or toxic substances. Damages include the actual clean-up costs incurred, and any legal or litigation costs and attorneys' fees related to the clean-up.

Contract Provisions That May Increase Exposure or Expenses

To reiterate, under the hazardous waste laws of the federal and state governments, owners and developers may be liable for damages and clean-up costs of property that is contaminated or may be contaminated with hazardous or toxic materials (Figure 5-1). Not surprisingly owners and developers sometimes attempt to shift this risk of liability to builders engaged in construction on the property. Builders hired to construct new homes on the owners' or developers' lots and remodelers hired to remodel or renovate existing structures must be alert to those contract provisions that may expose them to unwarranted liability (Figure 5-3).

When a new home or major addition is under construction, a common environmental concern is the discovery of an underground storage tank or some form of contaminated ground soil. If the builder's or remodeler's contract does not anticipate such an occurrence the builder or remodeler may find him- or herself responsible for one or more of the following actions:

- notifying the appropriate agencies of the environmental problem
- obtaining specialized permits
- removing the storage tank or hazardous material at his or her own expense or arranging for a third party to do so, also at the builder's expense

In the event an environmental problem arises, several contract provisions may expose the builder or remodeler to increased liability, or they may reduce the profitability of the project.

Figure 5-3. Contract Provisions That May Expose a Builder or Remodeler to Increased Environmental Liability

- Requirement for compliance with all laws and regulations
- Time extension
- No damages for delay
- Changed conditions and differing site conditions
- Change orders for environmental work
- Site inspection
- Builder's right to suspend performance
- Permission for the owner to employ independent contractors that requires the builder or remodeler to coordinate its work with the independent contractors at no increase in price

For example, a construction contract or remodeling contract might require the builder or remodeler to comply with all federal, state, and local laws and regulations. Such a clause might read: "The project will be completed in compliance with all laws, ordinances, rules, and regulations of the applicable governmental authorities." Builders and remodelers should delete the clause where possible, or they should limit their obligation to comply with laws and regulations to those which govern the means, method, and manner in which they perform their work.

Another problem clause involves concealed site conditions. As discussed in Chapters 2 and 3, unexpected or concealed site conditions can be expensive to work around. A standard contract (a) usually includes language contemplating the effect of unforeseen or concealed conditions and (b) may provide for notification, work changes, or contract modification procedures upon discovery of such a problem. Alternatively a contract could require the owner to pay the extra costs incurred, plus profit at a specified percentage of the cost.

If a builder or remodeler is not adequately licensed or otherwise equipped to remove hazardous waste, this obligation can be expensive and delay other aspects of the project. For this reason the contract ought to provide that the builder or remodeler has the option to cancel the contract upon the discovery of asbestos, polychlorinated biphenyls (PCBs), petroleum, hazardous waste, or other toxic substances. The builder or remodeler also may want to incorporate his or her own definition of hazardous waste in the contract. Alternatively the contract could provide that the owner is responsible for abatement and all costs associated with it.

Radon Liability

All parties to a real estate transaction or remodeling project should be concerned about the problem of naturally occurring radioactive radon. Radon is a naturally occurring gas that is caused by the radioactive decay of the element radium. No one can be blamed for its existence. But the lack of blame in the everyday sense of the word carries little weight in today's courtroom.

Although radon can be found in every home in the country and in the outside air, the Environmental Protection Agency (EPA) considers it a serious threat to health.[50] The only known health consequence of radon is an increased risk of developing lung cancer. But the extent of this risk posed by normal indoor levels of radon is disputed. Debating the health-risk and policy questions surrounding radon is beyond the scope of this book, but builders and remodelers should understand that they can be sued by home buyers and homeowners if elevated levels of radon are found in a home. This section discusses ways builders and remodelers can deal with potential radon liability.

Claims against builders and remodelers arising out of high radon levels found in homes can be brought under many different legal theories, including breach of express and implied warranties, negligence, strict liability, fraud and misrepresentation, and a host of other theories are based on state law.

With all of these theories, home buyers and homeowners may have difficulty proving in court that the radon in a particular home caused a homeowner's specific personal injury. However a court could accept merely the increased risk to health posed by radon as the legal injury even if the homeowner exhibits no actual injury. In addition, in some states homeowners have no need to show physical injury to recover damages for mental anguish and emotional distress.

Readers should consult local attorneys (a) for an explanation of the specific laws in their jurisdictions and (b) to determine whether the suggested contract language in this chapter addresses their specific state laws.

A buyer's or owner's own negligence that contributes to the injury, such as smoking or failure to maintain a radon mitigation system, may prevent or limit a builder's or remodeler's liability. Plaintiffs must also comply with applicable statutes of limitations, the time within which someone must file a suit. Both sides should recognize that such litigation will be expensive and unpredictable. The cost of correcting a radon problem in most homes ($300 to $2,500) is small in comparison to the expense of a lawsuit.

Reducing the Legal Risks

Although considerable potential radon liability exists, builders can take the steps listed in Figure 5-4 to reduce the risks.

Builders and remodelers who have provided specific warnings or notices in their contracts can control liability for a buyer's or homeowner's claim of fraud or misrepresentation for failure to warn about unsafe conditions. However some buyers may not enter into a contract if they see such language; they may wrongly assume that, because the contract has a radon notice, the house is "hot," while the builder down the street who does not have a radon notice must have a safe house. The builder must develop an effective notice that does not scare off buyers.

Remodelers must also educate homeowners about radon. The EPA provides this warning regarding radon and renovation:

> If you are planning any major structural renovation, such as converting an unfinished basement area into living space, it is especially important to test the area for radon before you begin the renovation. If your test results indicate a radon problem, radon-resistant techniques can be inexpensively included as part of the renovation. Because major renovations can change the level of radon in any home, always test again after work is completed.[51]

Though providing a radon notice provision in a contract is the prudent thing to do, a notice and the general concern over radon may breed demands for a radon inspection and sales contingency clause. If build-

Figure 5-4. Steps for Reducing Radon Liability Risks

- Builders and remodelers should use construction methods that the technical experts say will help to prevent radon from entering homes. All parties should know, however, that though it has produced some good preliminary research, the industry is at the beginning of a learning curve on this issue. No one knows exactly what works, and no construction technique is foolproof. Some techniques may add significant cost to the price of a home or a remodeling project—which is no small matter when extrapolated to the national housing and remodeling markets. In addition most of the techniques require homeowner maintenance to work properly.
- Builders and remodelers should contact the Technology and Codes Department of the National Association of Home Builders or the Environmental Protection Agency for the latest guidance on new home radon construction techniques.
- Builders, remodelers, and developers should include a specific notice regarding radon or indoor air quality in their contracts. An effective notice should (a) eliminate potential fraud and misrepresentation claims and (b) prevent potential negligence suits based on a failure to warn. However the contract for a real estate transaction should not contain a contingency based on short-term (3 to 7 days) radon test results (see Radon and Real Estate Transactions earlier in this chapter).
- Builders and remodelers should not give any warranties that would cover radon. No builder, remodeler, or seller of real estate can guarantee that a certain home will be safe from naturally occurring radon because the actual level of radon in a specific home depends on several factors that have nothing to do with the construction methods used, such as the occupants' living habits and weather conditions.
- An express warranty should be quite specific. It should enumerate each item or condition that the builder or remodeler intends to warrant. General or broad terms (as in any contract provision) should be avoided.
- Builders and remodelers should disclaim or waive all warranties, expressed or implied, that are not specifically enumerated in the builder's or remodeler's limited express warranty. With regard to new home construction, some states will not enforce this disclaimer as to implied warranties. About 20 states have specifically allowed the disclaimer. The trend toward allowing the disclaimer is increasing with the recognition of the harshness of the implied warranty rule on a builder or remodeler. Alternatively the warranty could be limited in duration to the term of the express warranty. The disclaimer language should be simple, clear, and conspicuous (for example, in bold type or in a box).
- The important point with all warranty provisions is for builders and remodelers to be sincere with potential buyers or owners, to let them know exactly what is warranted and what is not. The warranty should not leave any room for interpretation by either party or by a judge or a jury.
- Builders and remodelers should negotiate an indemnification and hold harmless clause in their contracts with foundation and ventilation subcontractors and design professionals in order to pass on any remaining liability to more responsible parties.

ers and buyers understand how radon enters a building and the problems with short-term radon tests, however, they should conclude that short-term inspection and contract contingencies are not in the parties' best interests. Therefore the parties to a real estate contract should not allow the sale to be contingent on short-term (3 to 7 days) radon test results.

The EPA's evaluation of radon levels and associated health risks is based on the average annual level of radon in a home and a lifetime (70 years) of exposure to those levels. In most cases a short-term screening measurement of radon levels is not a reliable measure of the average annual level.

Short-term tests cannot provide accurate estimates of long-term health risks because radon levels fluctuate daily, weekly, and seasonally, sometimes as much as

10-fold. Therefore short-term test results indicate only whether longer-term (3 months to 1 year) testing should be conducted.

According to the EPA, a short-term test that reveals a radon level of 4 picoCuries per liter (pCi/L) or higher (or 0.02 working levels or higher) means that the occupant should conduct either a long-term test within a year or a second short-term test (to determine the annual average level and the risk to health). Why should anyone (seller, buyer, agent, or lender) base a decision about a real estate transaction on information that provides no useful estimate of the potential risk to the occupants' health? Moreover a testing contingency in a contract provides no additional liability protection for any of the parties, assuming an effective notice is provided as explained above. However the issues of contract contingencies and radon inspections become consumer relations and business judgment decisions.

The EPA recommends that homeowners take remedial action within several weeks when the average of their two short-term screening tests is 4 Pci/L or higher. Therefore a contingency could be given to buyers for test results at these high levels. Until measurement methods progress to the point that accurate health risk estimates can be made from short-term tests, however, contract contingencies that are based on the results of a short-term screening test are unwise for all concerned.

Soil tests designed to predict radon levels that could occur in a building constructed on the site also are not recommended. A soil test that shows a low radon gas level does not indicate with a reasonable degree of assurance (from a liability perspective) that homes built on such soils will have low radon levels. Moreover a proper site survey, requiring multiple soil test samples, can be more expensive than the cost of construction techniques to prevent or mitigate radon. Before committing to soil tests, builders and remodelers should compare the cost and usefulness of any soil test proposals with the cost and benefits of radon prevention techniques that can be undertaken during construction of homes.

An escrow provision usually does not help the buyer because, if the market will bear it, the seller can simply inflate the price of the house to cover the cost of the escrow. The EPA estimates that nearly 1 out of 15 homes in the United States has an elevated level of radon. Given these numbers, over 90 percent of all buyers would pay for something they do not need if escrows were routinely required.

Lenders and others need not be as concerned with radon as they should be with potential liability for asbestos contamination in a building or with hazardous waste clean-up liability under the federal and state Superfund laws. Even high levels of radon can be reduced at a relatively low cost, unlike these other pollutants. The worst-case cost estimate for radon mitigation for a home is about $2,500. This amount is a substantial sum for many homeowners, but the average is more likely to be between $300 and $500. High radon levels do not lower the value of property; nor do the potential costs of mitigation create any real potential for default on a mortgage in comparison with the price of a house.

Radon Contract Provisions

An attorney may advise not including one or more of these provisions or parts of them in the contract for various reasons. An attorney also may advise combining various parts of the three provisions into one. Therefore builders and remodelers should not use these sample contract provisions without consulting an attorney experienced in construction contract law because of the differences in state laws.

Option 1—Radon Notice and Disclaimer— The U.S. Environmental Protection Agency, the U.S. Department of Health and Human Services, the U.S. Public Health Service, and the __(name the state department of health or environmental protection)__ have expressed concern over the presence of radon gas in homes. Prolonged exposure to high levels of indoor radon or its progeny may affect the health of residents. Although such conditions may exist, the __[builder or remodeler]__ has made no investigation to determine whether radon gas or other environmental pollutants are present in the home or affecting the premises. The builder has made no analysis or verification of the extent of the environmental or health hazard, if any, that may affect the premises or residents.

The __[builder or remodeler]__ makes no representation or warranty as to (a) the presence or lack of radon or hazardous environmental condition nor (b) the effect of radon or any such condition on the premises or residents.

The [builder or remodeler] further disclaims and [buyer or owner] waives—unless otherwise expressly provided for in the [builder's or remodeler's] limited warranty—all warranties, express or implied, including but not limited to the warranties of good workmanship, habitability, merchantability, and fitness of purpose and including any warranties that could be construed to cover the presence of radon or other environmental pollutants. The only warranties the [builder or remodeler] provides to the buyer are those contained in the limited warranty.

____
(buyers initials)

Option 2—Radon Notice—Radon is a naturally occurring gas that is caused by the radioactive decay of the element radium. Because radium is contained in the earth's crust and dissolves readily in water, radon can be found virtually everywhere. Preliminary studies suggest that prolonged indoor exposure to high levels of radon gas may result in adverse health consequences.

This notice serves to advise the [buyer or owner] that above-average levels of radon gas may accumulate in any home regardless of the type of home or who builds it. Specific radon levels depend on an array of factors, including—

- site-specific variables such as soil type and conditions, wind and climate conditions, geology and groundwater
- building-specific variables such as construction materials and techniques; age of the structure; heating, ventilating, and air-conditioning systems; and occupant upkeep and living habits

In addition, indoor radon levels can fluctuate on a yearly, seasonal, or even daily basis.

Because of the multitude of factors involved, predicting whether a residence may be subject to high radon levels is difficult, unless long-term tests (3 months to 1 year) to determine actual radon concentrations are conducted on a completed structure. The U.S. Environmental Protection Agency (EPA) and the U.S. Department of Energy are conducting extensive investigations regarding this matter.

The EPA has issued testing protocols that radon testing firms and homeowners should follow when testing homes. [home buyers and homeowners] may wish to make arrangements to test their homes for radon. Costs for radon test kits vary. EPA publishes a list of EPA-approved suppliers of such test kits. If a [home buyer or homeowner] decides to measure the level of radon gas in a home and if the tests detect an elevated level, the [home buyer or the homeowner] may wish to take steps to conduct longer-term testing or to reduce the level detected.

For remodelers add the following sample language:

The EPA's main public information pamphlet on radon, *A Citizen's Guide to Radon: What It Is and What To Do About It* (second edition), which the owner acknowledges receiving from the remodeler (_________ buyer's initials), provides that major renovations can change the level of radon in any home. Consequently the EPA advises testing for radon before renovation begins and after work is completed. The owner hereby releases the remodeler from all claims, losses, or demands (including personal or bodily injuries), and all of the consequences thereof, regardless if presently known, that may arise from the presence of radon in any structure on the property. The buyers further agree that they shall not withhold from the remodeler any payment or any portion thereof of the contract price for any reason, including radon test results or the presence of radon in the home.

[The remodeler disclaims and owner waives—unless otherwise expressly provided for in the remodeler's limited warranty—any warranties that could be construed to cover the presence of radon or other environmental pollutants. The only warranties the remodeler provides to the owner are those contained in the limited warranty.]

If the builder or remodeler has used sub-slab ventilation, the contract can state here:

One step the [home buyer or homeowner] may wish to take is to install (or have installed) a small ventilation fan as recommended by EPA to vent gas from beneath the slab of [the home or addition]. The [builder or remodeler] has prepared this [home or addition] for the installation of a sub-slab ventilation system. The [buyer or owner] may wish to contact EPA or the state's environmental protection office for further information on whether to install such a fan and on the choice of a fan. (EPA Regional Office address and phone number and the address and phone number of the applicable state office).

Alternatively the builder or remodeler may offer to test the home, at the buyer's expense, by stating:

At the act of sale or at closing, the [builder or remodeler] shall deliver to the [buyer or owner] a certificate from a firm that participates in the EPA's Radon Measurement Proficiency Program indicating what level of radon was present in the home at the time of the screening. The [buyer or owner] acknowledges that the [builder or remodeler] has not independently gathered any of the information that is contained within the certificate and that the [builder or remodeler] is merely providing the [buyer or the owner] with information that has been made available to it by an independent firm. The [buyer or the owner] further acknowledges that the [builder or remodeler] is not warranting the accuracy of the screening or the information contained within the certificate. Any rescreening of the residence or update of the certificate shall be performed by the [buyer or the owner] at the [buyer's or owner's] sole expense. (See also the no contingency clause that follows this one.)

The [buyer or the owner] acknowledges that the [builder or remodeler] does not claim or possess any special expertise in the measurement or reduction of radon, nor does [builder or remodeler] provide any advice to the [buyer or the owner] as to acceptable levels or possible health hazards of the gas. The buyers or owners agree that they shall not withhold from the builder or remodeler any payment or any portion of the contract price for any reason, including radon test results or the presence of radon in the home.

If the builder or the remodeler uses radon construction techniques, the contract may state:

The [builder or the remodeler] has relied on the expertise of the government agencies that have been studying the radon issue, and the [builder or remodeler] has constructed the home with the following techniques, as recommended by *Radon Reduction in New Construction: An Interim Guide*[52] (Describe construction techniques such as sub-slab ventilation, extra caulking, and others.) Because no construction technique is known to be foolproof, the [builder or the remodeler] assumes no responsibility for the operation, maintenance, or effectiveness of sub-slab ventilation systems or any other devices or methods intended for the reduction of radon gas levels.

Option 3—Contract Not Contingent on Radon—The U.S. Environmental Protection Agency (EPA) does not have mandatory regulations regarding radon levels in homes. However the EPA has issued *A Citizen's Guide to Radon*, (which the buyer acknowledges receiving from the builder [____________].)
buyer's initials

This guide explains what the owner of a home should do upon receiving the results of a radon test.

Radon is measured in terms of picoCuries per liter (pCi/L). The EPA discussion of pCi/L, its suggested actions, and its estimates of risk to health are based on the average annual level of radon in a home. In most cases a short-term (3 to 7 days) test of radon levels is not a reliable measure of the average annual level.

Therefore the buyers agree that this contract is not contingent on radon testing results or the presence or lack of radon in the home. The buyers further agree that they shall not extend the closing date or withhold from the builder any payment or any portion thereof of the contract price for any reason, including radon test results or the presence of radon in the home. By closing the purchase of the property, the buyer will be deemed to have released the builder from any and all claims and liabilities relating to or arising from the presence of radon or other environmental hazards on the property and from any and all responsibility for mitigating any elevated levels of radon gas or other environmental hazards that may be discovered on the property.

Asbestos Liability

Remodelers face several environmental risks that are not normally encountered in new home construction. Asbestos, one of the most common of these risks, is the name for a group of fibrous minerals found in rocks and soil. Because these minerals are heat-resistant and durable, asbestos has been used in construction for fire-proofing, insulation, and other building materials (Figure 5-5). Asbestos-related diseases are caused by asbestos fibers (detached from asbestos-containing products) that become suspended in the air and are inhaled or ingested. Asbestos may also be present in the soil and is sometimes encountered during excavation.

Asbestos and asbestos-containing materials (ACMs) generally fall into two broad categories—friable and nonfriable. When friable ACMs are dry, hand pressure can crumble, pulverize, or reduce them to powder and cause them to emit small fibers into the atmosphere. ACMs are nonfriable so long as the surface is stable, undamaged, and well sealed against the release of their fibers. Building materials that are ordinarily nonfriable may become friable and release asbestos fibers if they are subjected to sawing, sanding, drilling, grinding, or other similar treatment during renovation. These nor-

Figure 5-5. Asbestos Warning Signs

- Vinyl floor tiles
- Vinyl floor sheeting
- Vinyl-asbestos ceiling tiles
- Vinyl wallpaper
- Asbestos shingles and siding
- Pipe wrap used to insulate pipes, ducts, and furnaces
- Insulation sprayed behind walls and into ceilings for fireproofing, insulation, and soundproofing
- Heating and ventilation equipment
- Putty used at elbow and T-fittings and for flexible joints on heating equipment
- Drywall compound

mally nonfriable products include resilient floor coverings and asbestos roofing products.

A remodeler working in an existing home or a public or commercial building may encounter ACMs. Because asbestos and ACMs pose a serious health threat to remodelers, their workers, the owner and the residents of the house or building, and visitors to the home or building, ACMs should not be disturbed unless a remodeler is properly trained and, where necessary, certified or licensed as an abatement contractor.

In addition to their potential health risks, asbestos and ACMs also should concern remodelers because they present two liability problems: (1) federal, state, and local requirements and regulations and (2) contractual problems.

Federal, State, and Local Requirements and Regulations

Remodelers must comply with the regulation of asbestos and ACMs by federal, state, and local authorities. Failure to observe the laws that govern the use and handling of asbestos and ACMs may expose remodelers to great financial risk.

Four federal agencies regulate the use and handling of asbestos and ACMs: the EPA; the U.S. Department of Transportation (DOT); the U.S. Department of Labor, through the Occupational Safety and Health Administration (OSHA); and the Consumer Product Safety Commission (CPSC).

A number of states and local municipalities also regulate the use and handling of asbestos and ACMs. Other states have general laws pertaining to hazardous materials or wastes that may apply to asbestos. Remodelers need to be sure that they are familiar with any applicable state and local laws before they begin a renovation project.

The EPA regulations affect the remodeler-owner relationship most significantly. They apply to any "institutional, commercial, public, industrial, or residential structure, installation, or building (including any structure, installation, or building containing condominiums or individual dwelling units operated as a residential cooperative, but excluding residential buildings having four or fewer dwelling units.)"[53] Under the EPA regulations, before any of these structures are demolished or renovated, they must be inspected for the presence of asbestos. For renovations, if asbestos exceeds certain minimums, notification and removal requirements apply. All demolitions involving ACMs also must be reported to EPA, regardless of the quantity of ACMs involved.

The transportation of asbestos is regulated by DOT because it classifies asbestos as a hazardous substance. OSHA regulations protect the health and safety of workers involved with ACMs at worksites. CPSC provides educational material to the public regarding the dangers of ACMs, and it bans consumer products containing asbestos.

Contractual Problems

Because remodelers do not want to assume risk inadvertently, they must be sure that they are not bound by contract to remove or arrange for the removal of any ACMs just as they would do for any other toxic or hazardous materials or waste. If a site investigation, inspection, or other clause is drafted too broadly, it may expose the remodeler to unwarranted liability. Such a clause might read: "By executing the contract, the remodeler confirms that it has visited the site and familiarized itself with the work and the conditions under which the work is to be performed." Based on this language remodelers might find themselves responsible for removing or stabilizing any ACM they discover.

The remodeler can substantially reduce the risk described above by including language in the contract that expressly eliminates removing or stabilizing asbestos from the scope of the work. For example, remodelers may modify the language that appears above as follows:

> However the remodeler has not analyzed or verified the extent of the environmental or health hazards, if any, that may affect the residents of the premises. And the remodeler shall not be responsible for the detection, treatment, encapsulation, enclosure, or removal of any asbestos or any other hazardous material.

Statutes often require that such disclaimers be set in 10-point bold type or the computer equivalent of that size and density. **This sentence is set in 10-point bold type.**

In addition the contract should (a) address the costs likely to result from delayed completion of the project while the asbestos is being removed and (b) specifically provide for reimbursement for any additional cost caused by the discovery and abatement of ACM. The owner also should agree to indemnify the remodeler for costs incurred by the remodeler as a result of environmental problems associated with the property, for example, accidental contamination.

> The remodeler has made no analysis or verification of the extent of the environmental or health hazard, if any, that may affect the residents of the premises. The remodeler shall not be responsible for the detection, treatment, encapsulation, enclosure, or removal of any asbestos or other hazardous materials determined to be present at the site. The owner will be responsible for any treatment, encapsulation, enclosure, and removal of all hazardous materials including all associated expenses. Should the remodeler encounter materials on the site reasonably believed to be asbestos (or other hazardous waste), the remodeler shall have the right to stop work and remove its employees from the project until the nature of the substances has been determined and, if necessary, the substances have been removed or made harmless. The remodeler shall not be required to return to the site until the owner provides him or her with evidence that the asbestos has been removed or made harmless by a licensed asbestos abatement contractor.
>
> In the event work is suspended while asbestos is removed or stabilized, the remodeler shall be reimbursed for any additional cost resulting from the discovery of the asbestos and the resulting delay. Reimbursable costs shall include, but not be limited to, increased labor or material costs, increased finance costs, additional overhead costs, and start-up costs. To the fullest extent permitted by law, the owner shall indemnify and hold the remodeler harmless from and against all claims, costs, losses, damages, and expenses, including attorney's fees, arising from or involving such hazardous materials.

> **The remodeler disclaims and owner waives all warranties that could be construed to cover the presence of asbestos or other environmental pollutants. Should the owner elect not to proceed with this project because of asbestos or some other environmental health hazard, the remodeler shall be entitled to terminate the contract and shall be entitled to receive money for all unpaid costs, fees, and expenses, including the prorated cost of overhead expenses, earned to the time of termination, as well as a prorated percentage of the remodeler's total anticipated profits.**

Inclusion of a clause like the immediately preceding one may require modification of other contract provisions. For example, if the contract contains a no-damages-for-delay clause, arguably that provision would conflict with language in the preceding clause that requires the owner to reimburse the remodeler for any additional cost caused by the discovery of the asbestos and the resulting delay.

This problem could be avoided by providing that the remodeler will not be entitled to damages for delay except where the delay results from the discovery of asbestos or other hazardous material.

Typically builders and remodelers do not want to include a no-damages-for delay clause because such a clause provides that if the builder or remodeler is delayed, no damages will be paid as a result of the delay. They and their attorneys need to watch for such a clause, however, because often buyers' or owners' attorneys want to insert one in their clients' contracts to purchase or remodel a house.

Lead

Remodelers face another environmental risk from lead. Though lead comes from various sources, two of the most hazardous forms of lead are lead paint and lead in dust and soil. The federal Lead-Based Paint Hazard Reduction Act of 1992 (Title X) contains provisions that remodelers should be aware of. The law required OSHA to issue regulations governing occupational exposure to lead in the construction industry.

The law also requires remodelers to issue a lead hazard information pamphlet to owners and occupants of houses on which they work. These requirements as well as potential liability issues are explained below.

OSHA Interim Final Lead in Construction Standard

OSHA issued its Interim Final Lead in Construction Standard in 1933. The Standard reduced the Permissible Exposure Limit (PEL) for lead for construction workers from 200 $\mu g/m^3$ to 50 $\mu g/m^3$ and the action level (AL) for medical surveillance and training to 30 $\mu g/m^3$. The rule requires that the air on a remodeling job for lead be monitored frequently for lead. If activities generate lead dust in the air below 30 $\mu g/m^3$, special worker protection requirements are not required. Respirators, protective clothing, blood testing, training, a written compliance plan, and recordkeeping are required above the PEL. Workers must wear respirators and protective clothing until laboratory test results show that the air lead level for the activity is below 50 $\mu g/m^3$. For more complete information, see *What Remodelers Need to Know and Do About Lead.*[54]

Lead Abatement

Under Title X, lead abatement activities such as the permanent elimination of lead-based paint must be conducted by lead abatement professionals. These individuals must be trained and certified to do such work in accordance with state or federal certification requirements.

EPA is currently issuing certification and training requirements for lead-based paint activities such as abatement. Once the regulations are finalized, the states have 2 years to develop their own certification and training programs. Thus remodelers who perform lead-based paint activities such as abatement will have to become certified under federal or state law.

Abatement does not include renovation and remodeling or landscaping activities the primary intent of which is not to eliminate lead-based paint hazards permanently, but instead to repair, restore, or remodel a dwelling, even though these activities may incidently result in a reduction in lead-based paint.

Distribution of Lead Information Pamphlet

Title X requires that before a remodeler undertakes renovation and remodeling activities in housing built before 1978, the remodeler must provide the owner and occupant a lead information pamphlet developed by the EPA and the Department of Housing and Urban Development. Final rules implementing this requirement are expected in early 1996 and will probably take effect by mid-1996.

Remodelers will need to have owners or occupants sign a statement indicating that they (a) have received the lead information pamphlet and (b) are aware of hazards associated with renovating housing containing lead-based paint. This statement can be on a separate sheet or incorporated into the sales contract. Remodelers will need to keep records documenting compliance with the rule for a specific period of time.

Lead Liability

Few lawsuits involving lead have been filed against remodelers to date. However remodelers may be sued for exposing their workers, their clients, and their clients' children. Remodelers are especially subject to liability if children under 6 years remain in the home during remodeling because they may be exposed to lead dust and lead chips during this activity.

Remodelers may be held liable to their clients for (a) not meeting the terms of their contract (contract liability); (b) disregarding local, state, or federal laws (statutory liability), including OSHA regulations; and (c) not meeting standards expected of ordinary, reasonable people (common law). Tort liability arises from not meeting expected standards. If a remodeler knows that lead is present and knows what precautions he or she should take, but does not warn the client about the hazard nor take appropriate action, the remodeler may be liable for negligence.

In many jurisdictions lack of knowledge will not reduce a remodeler's liability. In addition the courts may require a higher standard of care than legislatures and local governments.

To minimize their liability, remodelers should reduce and contain lead dust while working and clean thoroughly when the job is complete. They may also want to test the floor for lead dust with a lead test kit in the presence of the client and have the client sign a statement that he or she witnessed the test. For maximum protection, however, the remodeler should hire a testing company to take dust samples before and after the job for analysis in a laboratory to provide better evidence that remodeling did not increase dust levels.

If a remodeler decides to collect the wipe samples and send them to a laboratory for analysis, the remodeler will have a stronger defense if he or she is trained and certified to do so (see Figure 5-6 to reduce risks from remodeling).

Finally, if a potential client has children under 6 years old and a house with visible dust and/or peeling paint, a remodeler must be especially careful about cleanliness if he or she accepts the job. The children may already be poisoned, and the remodeler may be blamed for it after the job. A wise procedure is to ask the client before accepting a job whether the children have had blood tests for lead. If the client says no, the remodeler should use caution. If the client says the children's blood lead levels were well below 10ug/dl, the remodeler could ask for a copy of the test results.

Remodelers can protect themselves against possible lawsuits by employees who claim they were lead-poisoned (while working for the remodelers) by requiring all new employees to have a blood test to determine their lead exposure level before working for the remodelers. Remodelers should keep the results on file even after an employee leaves.

Figure 5-6. Steps for Reducing a Remodeler's Lead Liability Risks

- Provide owners with Environmental Protection Agency pamphlet advising them of health hazards associated with lead and warning them of possible hazards of lead dust generated during remodeling.
- Have the owner sign statement indicating that the owner has received the EPA pamphlet.
- Do not undertake lead abatement work without training and certification.
- Use wet sanding methods wherever feasible.
- Isolate any work areas with plastic sheeting where significant dust is expected, including floors, furniture, heating registers, and cold air returns.
- Use vacuum attachments on saws and sanders wherever possible.
- Wash and vacuum floors exposed to dust during remodeling. Be sure washing will not damage them. Use only high-efficiency (HEPA) filters.
- Vacuum floors before installing new carpeting. Use only high-efficiency (HEPA) filters.

Six • Warranties and Disclaimers

Express and Implied Warranties

To a large extent builders' and remodelers' liabilities for correcting problems after the work is completed depend on the warranties that they give to their customers. Every time builders or remodelers, their agents, and their employees make a promise to a buyer or an owner, they may be extending a warranty. This fact also applies to any representations made in advertisements, brochures, and correspondence. However warranties do not have to be in writing; they may also be spoken or implied. An "express warranty"—a warranty in words, whether spoken or written—is treated by the courts like any other agreement. Therefore, if a builder or remodeler breaches such a warranty, that act is like breaching a contract.

In addition to express warranties, courts have found "implied warranties" in contracts to construct new homes or to remodel existing ones. Even if builders or remodelers make no specific written or oral promises about the condition of the houses or their residential remodeling projects, many courts read into the contract a promise to provide a house that is reasonably free from defects in workmanship or materials and, in the case of a new home, that is habitable. In some states home improvement contracts may be covered by an implied warranty or a statutory express warranty. For example, the Indiana Home Improvement Warranty Act allows a remodeler to disclaim implied warranties in a home improvement contract only if the remodeler offers the owner the express warranties defined in the Act. (Ind. Code §24-5-11.5-1). These implied warranties are dangerous because the courts—not the parties to the contract—determine what is covered by the warranties.

The warranty of habitability or fitness requires that the new home be fit for its intended use—habitation. What constitutes habitability is a difficult question to answer. At a minimum, however, a new home must—

- substantially comply with all building and housing codes.
- keep out the elements and provide its inhabitants with a reasonably safe place to live, without fear of injury to person, health, safety, or property
- be structurally sound, and it should include sufficient heat, ventilation, and light; adequate plumbing and sanitation (including potable water), and proper security

Under the implied warranty of workmanship, the builder warrants that the home meets the standards of quality prevailing at the time and place of construction. The builder is not required to build a perfect home. However the courts have upheld implied warranty claims for an assortment of defects, including but not limited to shrinkage of exterior siding, cracking foundations, wet basements, leaky roofs, sagging roofs and floors, defective air-conditioning and heating systems, defective tank and drain field systems, and unusable water supplies.

Numerous states now extend the implied warranties to second, third, or later owners of homes, although those warranties are limited to latent defects (not discoverable upon reasonable inspection) that manifest themselves within a reasonable period of time after construction of the house. The length of time the original or subsequent buyer has to enforce the warranty varies from state to state, but a number of courts have established a standard of reasonableness. The age of the home, its maintenance, and the use to which it has been put are a few of the factors that a court considers. What is reasonable depends on the component part of the home involved. For example, a roof may be expected to have a longer useful life than a termite treatment.

Implied warranties add to the provisions of an express warranty. Thus a builder who offers an express warranty also may be bound by the implied warranties unless the builder has properly disclaimed the implied warranties.

Illustrative Case

Ferrell built a single-family residence in 1980 and sold the residence to a private party who subsequently resold the house in 1981 to the Bridges. The original sales contract provided for a "1-year Builder's Warranty," and it was transferred to the Bridges upon resale. Shortly after the Bridges took possession, they experienced structural problems with the house, and they sued Ferrell for breach of implied warranty of habitability, recovering $9,000.

Ferrell appealed on the grounds that the court erroneously instructed the jury that the implied warranty of habitability controlled over an express warranty agreed on by the parties. The Oklahoma Court of Appeals concluded that Ferrell was asking the court to instruct the jury that a written warranty agreed to by the parties for 1 year controlled over an implied warranty for a reasonable period of time. The Appeals Court rejected Ferrell's requested instruction as an incorrect statement of the law. The court held that, in the absence of an agreement to the contrary, the mere existence of an express warranty does not displace the obligations arising by operation of law under an implied warranty of habitability.[55]

In theory, because a warranty is like any other promise in a contract, a builder or remodeler should be able to write a contract disclaiming or limiting any warranties, even implied warranties. In an increasing number of states the courts have enforced contract provisions in which the parties agree to waive implied warranties. In other states, however, courts have held that (a) consumers need special protection when purchasing or remodeling their homes, which often are their most expensive investments, and (b) they must not be allowed to waive the implied warranties.

Before attempting to disclaim all implied warranties, builders and remodelers need to consider the following issues:

- The trend in the courts is toward including more items within the reach of the Magnuson-Moss Warranty Act, and if builders and remodelers offer a full express warranty, they cannot limit implied warranties under the Act (see the Magnuson-Moss Act described below).
- Some courts will not uphold a contract provision that is deemed unscrupulous or unfair to the consumer.
- From a marketing perspective some builders and remodelers have found that providing a warranty that restricts the duration of the implied warranties to the period of time covered by the express warranties makes better business sense than to eliminate all implied warranties. For example, the builder might offer the owner a 1-year builder-backed warranty and a 1-year implied warranty.

The warranty should conform to certain standards because the buyers' or owners' personal standards may differ from industry standards. If the parties do not spell out the construction standards, they run the risk of having an arbitrator or court decide for them. One source of industry standards is *Quality Standards for the Professional Remodeler*.[56] Alternately a builder or remodeler could develop his or her own standards.

Illustrative Case

The purchasers of a new home brought an action against the builder for breach of implied warranty of habitability. The builder argued that the purchasers waived their right to sue for breach of implied warranty of habitability by signing and initialing the agreement that provided for a homeowners' warranty in lieu of the implied warranty of habitability. The agreement provided that (a) the seller expressly disclaimed the implied warranty of habitability and (b), as a result of that disclaimer by the seller, the seller's sole warranty given to the purchaser was the homeowners' warranty. Finding for the builder, the court held that the evidence showed that the disclaimer language was brought to the purchasers' attention, that the consequences of the agreement were made known to them, and that they knowingly waived their rights to pursue an action against the builder for breach of the implied warranty of habitability.[57]

The Magnuson-Moss Warranty Act

The Federal Trade Commission (FTC), a government agency, regulates warranties on consumer products under the Magnuson-Moss Warranty Act.[58] The act does not require a builder or remodeler to give a warranty. It merely regulates the form of whatever warranty a builder may voluntarily decide to give. The act applies to builder- and remodeler-backed warranties and insured warranty programs. The act states that these warranties must meet certain disclosure requirements if they are to be valid. These requirements do not apply to new or remodeled houses themselves, but only to the following items that are installed or included in new or remodeled houses: appliances, manufactured equipment, and anything else defined as a consumer product under the act.

Builders' and remodelers' warranty provisions that cover consumer products must be written according to FTC regulations.

Builders and remodelers face three choices when considering their responsibilities under the Magnuson-Moss Act:

- Give no written warranty at all, which may not be practical in many markets.
- Exclude consumer products entirely from their warranty and, thereby, avoid triggering the Magnuson-Moss requirements. However builders and remodelers should keep in mind that they will be excluding an enormous variety of appliances and pieces of equipment.
- Conform their warranties to Magnuson-Moss guidelines.

Because the act applies only if a warranty covers consumer products, a warranty that excludes all consumer products from coverage is not affected by the act. Following are examples of language that a builder or remodeler might use to exclude consumer products from his or her written warranty:

1. "This warranty does not cover any appliance piece of equipment, or other item which is a consumer product for purposes of the Magnuson-Moss Warranty Act (15 U.S.C. Sec. 2301 through 2312.)"

2. "This warranty does not cover any appliance, piece of equipment, or other item in the home, which is a 'consumer product' for the purposes of the Magnuson-Moss Warranty Act (15 United States Code §2301 through 2312). The following are examples of 'consumer products,' *but . . . other items in the home . . . also [may be] consumer products:* (List the items from the FTC list)." [See the discussion of limited warranty agreements excluding items covered by the Magnuson-Moss Act later in this chapter.]"

The second example above offers more protection legally, but may be awkward to include in a brief warranty document.[59]

Excluding appliances and pieces of equipment from a builder's or remodeler's written warranty will prevent it from being subject to the FTC's requirements regarding its form and language. However, if the warranty exclusions leave nothing covered, the FTC could consider the warranty a sham under Section 5 of the Federal Trade Commission Act, which covers "unfair and deceptive acts and practices."[60]

Because the act pertains primarily to the form of a warranty rather than to a builder's or a remodeler's obligations, the easiest solution might be to write "the entire warranty in the form required by the act, even though doing so goes beyond what the law technically requires. . . . [Thus, the builder or remodeler can] provide a warranty on any appliances and pieces of equipment . . . [without deciding] precisely which are consumer products. . . . something no one has been able to explain adequately in any type of language."[61]

If any part of a builder's or remodeler's warranty is covered by Magnuson-Moss and the builder provides for alternative dispute resolution of claims, the warranty must follow the settlement procedures outlined in the act. The act authorizes the FTC to regulate devices under which warranty disputes between a builder and purchaser are submitted to a third party for mediation or arbitration (see Arbitration and Mediation in Chapter 2 for builders and in Chapter 3 for remodelers).

These informal devices are referred to in the act as "informal dispute settlement mechanisms." A builder is not required to provide such a mechanism. If he or she voluntarily provides for one in his or her warranty document, however, it must meet FTC standards. These rules are complicated and builders and remodelers are advised to review their warranty documents with their attorneys to assure that the warranty documents do not run afoul of the act.

The Magnuson-Moss Act does not allow builders or remodelers to disclaim implied warranties on consumer products if they give a full, written warranty (see Limited Warranties later in this chapter). If builders and remodelers follow the Magnuson-Moss Act guidelines for limited warranties, they are permitted to restrict the duration of the implied warranties (unless state law does not permit the restriction). The fairly simple requirements are designed merely to make warranties more understandable to consumers. Some courts are more likely to enforce a limited warranty if it is simple, clear, and easy for a home buyer or homeowner to understand even if the Magnuson-Moss Act does not specifically apply.

Under the FTC regulations, consumers must have access to consumer product warranties before a sale. Sellers can display the warranty near the warranted product or make it available upon request. Sellers who choose to make warranties available upon request must post signs to that effect, and the warranty document must include the information in Figure 6-1.

Figure 6-1. Federal Trade Regulation Requirements for Warranty Documents

- The name of the person who is receiving the warranty and whether it is transferable to subsequent purchasers
- Precisely what the warranty covers and what is excluded
- What the builder or the remodeler will do if a warranted defect or problem occurs and how much the consumer will be charged
- The duration of the warranty and when it begins
- The procedures a buyer or homeowner must follow to have a problem corrected
- Limitations on consequential or secondary results must stand out from the rest of the document, and they must be followed by these exact words: "Some states do not allow the exclusion or limitation of incidental or consequential damages, so the above limitation or exclusion may not apply to you."
- Any limitations on the duration of implied warranty rights also must appear conspicuously in the document, along with the statement: "Some states do not allow limitations on how long an implied warranty lasts, so this limitation may not apply to you."
- A provision stating: "This warranty gives you specific legal rights, and you may also have other rights which vary from state to state."

Full and Limited Warranty

The language of any written warranty that a builder provides to a buyer or that a remodeler provides to a homeowner may determine the extent of the builder's or remodeler's liability. Although the language is simple, it has legal effect.

The Magnuson-Moss Act distinguishes between full and limited warranties. A full warranty must- do the following:

- provide for remedies within a reasonable time (at no charge) for consumer products with defects, malfunctions, or for failure to conform to the warranty
- avoid imposing any limitation on the duration of implied warranties
- conspicuously state on the document itself whether consequential damages are limited or excluded
- permit the consumer to elect either a refund for or a free replacement of an item with a defect that cannot be repaired after a reasonable number of attempts

Any warranty that does not meet the minimum standards of a full warranty as outlined above is considered a limited warranty. A full warranty provides for replacement of the product warranted, so the consumer would receive a new one if anything goes wrong with the original product during the warranty period. Replacement of an entire house if something goes wrong with one part is preposterous, so warranties on houses are always limited warranties—providing for repair of specified items or redoing work rather than replacement.

As explained in the opening of this chapter, builders and remodelers should use a limited express warranty that restricts the duration of implied warranties. After evaluating the risks previously outlined in this chapter, builders and remodelers should consider totally disclaiming all implied warranties in those states that permit it.

The remainder of this chapter includes the following items:

- an alternative limited warranty agreement excluding items covered by the Magnuson-Moss Act (Figure 6-2 for builders and Figure 6-3 for remodelers)
- a limited warranty agreement conforming to the Magnuson-Moss Act (Figure 6-4 for either)
- a statement of nonwarrantable conditions (Figure 6-5)

Builders and remodelers should not use these provisions without consulting an attorney experienced in building industry warranties and revising them to fit the facts of each situation. They should study the preceding portion of this chapter to understand the significance of the language contained in these limited warranties. They also must coordinate the warranties they use with the sections of their contracts that cover (a) Inspections, Acceptance, and Possession and (b) Representations and Warranty in Chapter 2 for builders and Chapter 3 for remodelers.

Builders and remodelers who participate in insured warranty programs should redraft the sample language presented in this chapter or disregard it to avoid conflict with their insured warranty programs.

Figure 6-2. Builder's Sample Limited Warranty Agreement That Excludes Items Covered by the Magnuson-Moss Act

This limited warranty agreement is extended by (builder's name), (the builder), whose address is (builder's address), to (buyer's name), (the buyer), who is the original buyer of the property at the following address:[1]

____

____

____

____

____

This limited warranty excludes consequential damages, limits the duration of implied warranties, and provides for liquidated damages.

1. What is Covered by the Warranty?

The builder warrants that all construction related to the house substantially conforms with the plans and specifications and change orders for this job. The builder warrants that during the first 30 days after the buyer moves in, the builder will adjust or correct minor defects, omissions, or malfunctions, such as missing equipment or hardware; sticking doors, drawers, and windows; dripping faucets; and other minor malfunctions reported by the buyer upon inspection of the property.

Within one (1) year from the date of closing or occupancy by the buyer, whichever is first, the builder will repair or replace, at the builder's option, any latent defects in material or workmanship by the standards of construction relevant in (city, state). A latent defect is defined as one which was not apparent or ascertainable at the time of occupancy. The buyer agrees to accept a reasonable match in any repair or replacement in the event the original item is no longer available.

2. What is Not Covered?

This limited warranty does not cover the following items:

A. Damage resulting from fires, floods, storms, electrical malfunctions, accidents, or acts of God

B. Damage from alterations, misuse, or abuse of the covered items by any person

C. Damage resulting from the buyer's failure to observe any operating instructions furnished by the builder at the time of installation

D. Damage resulting from a malfunction of equipment or lines of the telephone, gas, power, or water companies

E. Any items listed as Nonwarrantable Conditions on the list that is incorporated in this contract; the buyer acknowledges receipt of the list of Nonwarrantable Conditions __________ (buyer's initials)

F. Any item furnished or installed by the buyer

G. Any appliance, piece of equipment, or other item that is a consumer product for the purposes of the Magnuson-Moss Warranty Act, 15 United States Code §2301 et seq., installed or included in the buyer's property.

The only warranties on items listed below are those that the manufacturer provides to the buyer:[2]

Appliances

Clothes dryer
Clothes washer
Dishwasher
Freezer
Garbage disposal
Ice maker
Kitchen center [a type of food processor]
Microwave
Oven and oven hood
Refrigerator
Range, stove, or cooktop
Trash compactor

Heating and Ventilation

Air-conditioning
Boiler
Electronic air cleaner
Exhaust fan
Furnace
Heat pump
Humidifier
Space heater
Thermostat

Mechanical and/or Electrical

Burglar alarm
Central vacuum system
Chimes
Electric meter
Fire alarm
Fire extinguisher
Garage door opener
Gas meter
Gas or electric barbecue grill
Intercom
Smoke detector
Water meter
Water pump

Plumbing

Garbage disposal
Sump pump
Water heater
Water softener
Whirlpool bath

The following items are not consumer products under the Magnuson-Moss Warranty Act when sold as part of a new home:

---

1. This form is designed for a single buyer. If more than one buyer is involved, the form should be adapted to accommodate the initials and signature of each of the buyers.

2. **Warning**—This list is not exclusive.

Figure 6-2. Builder's Sample Limited Warranty Agreement That Excludes Items Covered by the Magnuson-Moss Act (continued)

Heating and Ventilation

Duct	Register	Radiator

Mechanical and/or Electrical

Circuit breaker	Electrical panel box	Garage door
Electrical switch and outlet	Fuse	Light fixture
		Wiring

Plumbing

Bidet	Shower stall	faucet, trap, and drain)
Bathtub	Sprinkler head	Toilet
Laundry tray	Plumbing fittings (showerhead,	Vanity
Medicine cabinet		
Sink		

Miscellaneous Items

Cabinet	Floor covering (includes carpeting, linoleum, tile, parquet)	Shelving
Ceiling	Gutter	Shingles
Chimney and fireplace		Wall or wall covering
Door		Window
Fencing		

The following separate items of equipment are not consumer products under the Magnuson-Moss Act when sold as part of a condominium, cooperative, or similar multiple-family dwelling . . . [because] they are not normally used for "personal, family, or household purposes" within the meaning of the act:

Elevator
Emergency back-up generator
Institutional trash compactor
Fusible fire door closer
Master TV antenna
TV security monitor

[If the item has a function separate and apart from the house, it is likely to be considered a consumer product (such as a water heater, stove, or refrigerator), whereas other items (such as floorboards and trusses) are not.]

(1) The builder has made any such warranties available to the buyer for the buyer's inspection and the buyer acknowledges receipt of copies of any warranties requested. __________ (buyer's initials)

(2) The builder hereby assigns (to the extent that they are assignable) and conveys to the buyer all warranties provided to the builder on any manufactured items that have been installed or included in the buyer's property. The buyer accepts this assignment and acknowledges that the builder's only responsibility relating to such items is to lend assistance to the buyer in settling any claim resulting from the installation of these products. __________ (buyer's initials) __________ (builder's initials)

3. Remedies and Limitations

A. The buyer understands that the sole remedies under this limited warranty agreement are repair and replacement as set forth here. __________ (buyer's initials)

B. With respect to any claim whatsoever asserted by the buyer against the builder, the buyer understands that the buyer will have no right to recover or request compensation for, and the builder shall not be liable for—

(1) Incidental, consequential, secondary, or punitive damages
(2) Damages for aggravation, mental anguish, emotional distress, or pain and suffering
(3) Attorney's fees or costs __________ (buyer's initials)

C. The builder hereby limits the duration of all implied warranties, including the implied warranties of habitability, and workmanlike construction to one (1) year from the date of sale or the date of occupancy, whichever comes first. __________ (buyer's initials)

D. These limitations shall be enforceable to the extent permitted by law. [Some states do not allow the exclusion or limitation of incidental or consequential damages or the limitation of implied warranties, so the limitations or exclusions listed above may not apply in a particular location.]

[Alternative to 3C

[C. The buyer understands that no implied warranties whatsoever apply to the structure of the house and items that are functionally part of the house. The builder disclaims any implied warranties, including (but not limited to) warranties of habitability, workmanship, and materials to the extent allowed by law, and any implied warranty that exists despite this disclaimer is limited to a period of one (1) year. These limitations shall be enforceable to the extent permitted by the law. Some states do not allow limitations on how long an implied warranty lasts, so this limitation may not apply. __________ (buyer's initials)

Figure 6-2. Builder's Sample Limited Warranty Agreement That Excludes Items Covered by the Magnuson-Moss Act (continued)

[The buyer acknowledges acceptance of these limitations on the warranties offered by the builder in consideration for the limited warranty and the other provisions of the construction contract.]

D. Notwithstanding the provisions of this limited warranty agreement, if any liability arises on the part of the builder, the builder will pay the amount of actual provable damages arising from such liability, but the amount shall not exceed $________. This amount, fixed as liquidated damages and not as a penalty, shall be the builder's complete and exclusive amount of liability. The provisions of this paragraph apply if loss or damage results directly or indirectly to persons or property from the performance of, or failure to perform, obligations imposed by the construction contract or from negligence, active or otherwise, of the builder, the builder's agents, or employees.

The buyer (a) understands that this provision limits the damages for which the builder will be liable and (b) acknowledges acceptance of this liquidated damages provision in consideration for the limited warranties provided by the builder and the other provisions of the construction contract. Therefore the buyer agrees to this liquidated damages clause if, notwithstanding the provisions of this limited warranty, liability should arise on the part of the builder.

E. This warranty is personal to the original buyer and does not run with the property or the items contained in the house. The original buyer may not assign, transfer, or convey this warranty without the prior written consent of the builder.

____
(buyer's initials)

4. How to Obtain Service

If a problem develops during the warranty period, the buyer should notify the builder in writing at the address given above of the specific problem. The written statement of the problem should include the buyer's name, address, telephone number, and a description of the nature of the problem. The builder will begin performing the obligations under this warranty within a reasonable time of the builder's receipt of such a request and will diligently pursue these obligations.

Repair work will be done during the builder's normal working hours except where delay will cause additional damage. The buyer agrees to provide the builder or builder's representative access to the house. The buyer also agrees to provide the presence (during the work) of a responsible adult with the authority to approve the repair and sign an acceptance of repair ticket upon completion of the repair.

5. Specific Legal Rights

This limited warranty gives the buyer specific legal rights, and the buyer may also have other rights that vary from state to state.

6. Where to Get Help

If the buyer wants help or information concerning this warranty, the buyer should contact the builder.

7. The Only Warranty Given by the Builder

The buyer acknowledges (a) that __[he or she]__ has thoroughly examined the property to be conveyed, (b) the buyer has read and understands the limited warranty, and (c) the builder has made no guarantees, warranties, understandings, nor representations (nor have any been made by any representatives of the builder) that are not set forth in this document.

I acknowledge having read, understood, and received a copy of this limited warranty agreement.

________________ (buyer)

Date ________________

________________ (builder)

By ________________

Title ________________

Date ________________

Figure 6-3. Remodeler's Sample Limited Warranty Agreement That Excludes Items Covered by the Magnuson-Moss Act

This limited warranty agreement is extended by _(remodeler's name)_ (the remodeler), whose address is _(remodeler's address)_, to _(owner's name)_ (the owner) of the property at the following address (house):[1]

____

____

____

____

This limited warranty excludes consequential damages, limits the duration of implied warranties, and provides for liquidated damages.

1. What is Covered by the Warranty?

The remodeler warrants that all construction related to the _[remodeling, renovation, rehabilitation, restoration, or addition]_ substantially conforms with the plans and specifications and change orders for this job. The remodeler warrants that during the first thirty (30) days after the owner occupies the remodeled space, the remodeler will adjust or correct minor defects, omissions, or malfunctions, such as missing equipment or hardware; sticking doors, drawers, and windows; dripping faucets; and other minor malfunctions reported by the owner upon inspection of the _[remodeled, renovated, rehabilitated, restored, or added]_ space.

Within one (1) year from the date of substantial completion or use of the _[remodeled, renovated, rehabilitated, restored, or added]_ space by the owner, whichever is first, the remodeler will repair or replace, at the remodeler's option, any latent defects in material or workmanship by the standards of construction relevant in _(city, state)_. A latent defect is defined as one which was not apparent or ascertainable at the time of occupancy. The owner agrees to accept a reasonable match in any repair or replacement in the event the original item is no longer available.

2. What is Not Covered

This limited warranty does not cover the following items:

A. Damage resulting from fires, floods, storms, electrical malfunctions, accidents, or acts of God
B. Damage from alterations, misuse, or abuse of the covered items by any person
C. Damage resulting from the owner's failure to observe any operating instructions furnished by the remodeler at the time of installation
D. Damage resulting from a malfunction of equipment or lines of the telephone, gas, power, or water companies
E. Any items listed as nonwarrantable conditions on the list that is incorporated into this contract (The owner acknowledges receipt of the list of nonwarrantable conditions.) _________ (owner's initials)
F. Any item furnished or installed by the owner
G. Any appliance, piece of equipment, or other item that is a consumer product for the purposes of the Magnuson-Moss Warranty Act, 15 United States Code §2301 et seq., installed or included in the owner's property

The only warranties of items listed below are those that the manufacturer provides to the owner:[3]

Appliances

Clothes dryer	Kitchen center (a type of food processor)	Refrigerator
Clothes washer	Microwave	Range, stove, or cooktop
Dishwasher	Oven and oven hood	Trash compactor
Freezer		
Garbage disposal		
Ice maker		

Heating and Ventilation

Air-conditioning	Exhaust fan	Space heater
Boiler	Furnace	Thermostat
Electronic air cleaner	Heat pump	
	Humidifier	

Mechanical and/or Electrical

Burglar alarm	Fire extinguisher	Smoke detector
Central vacuum system	Garage door opener	Water meter
Chimes	Gas meter	Water pump
Electric meter	Gas or electric barbecue grill	
Fire alarm	Intercom	

Plumbing

Garbage disposal	Water heater	Whirlpool bath[2]
Sump pump	Water softener	

1. This form is designed for a single owner. If more than one owner is involved, it should be adapted to accommodate the initials and signature of each of the owners.

2. Remodelers should exclude from this list any items that are not applicable to the job for which the warranty is being provided. For instance, if the job did not involve any kitchen appliances, the remodeler need not include them in the list.

3. **Warning**—This list is not exclusive.

Figure 6-3. Remodeler's Sample Limited Warranty Agreement That Excludes Items Covered by the Magnuson-Moss Act (continued)

The following items are not consumer products under the Magnuson-Moss Warranty Act when sold as part of a new home:

Heating and Ventilation

Duct	Register	Radiator

Mechanical and/or Electrical

Circuit breaker	Electrical panel box	Garage door
Electrical switch and outlet	Fuse	Light fixture
		Wiring

Plumbing

Bidet	Plumbing fittings (showerhead, faucet, trap, escutcheon [flange around a pipe or fitting], and drain)	Toilet
Bathtub		Vanity
Medicine cabinet		
Sink		
Shower stall		
Sprinkler head		

Miscellaneous Items

Cabinet	Floor covering (includes carpeting, linoleum, tile, parquet)	Shelving
Ceiling	Gutter	Shingles
Chimney and fireplace		Wall or wall covering
Door		Window
Fencing		

The following separate items of equipment are not consumer products under the Magnuson-Moss Act when sold as part of a condominium, cooperative, or similar multiple-family dwelling . . . [because] they are not normally used for "personal, family, or household purposes" within the meaning of the act:

Elevator
Emergency back-up generator
Institutional trash compactor
Fusible fire door closer
Master TV antenna
TV security monitor

(1) The remodeler has made any such warranties available to the owner for the owner's inspection and the owner acknowledges receipt of copies of any warranties requested. ___________
(owner's initials)

(2) The remodeler hereby assigns (to the extent that they are assignable) and conveys to the owner all warranties provided to the remodeler on any manufactured items that have been installed or included in the owner's property. The owner accepts this assignment and acknowledges that the remodeler's only responsibility relating to such items is to lend assistance to the owner in settling any claim resulting from the installation of these products. ___________ ___________
(owner's initials) (remodeler's initials)

3. Remedies and Limitations

A. The owner understands that the sole remedies under this limited warranty agreement are repair and replacement as set forth here. ___________
(owner's initials)

B. With respect to any claim whatsoever asserted by the owner against the remodeler, the owner understands that the owner will have no right to recover or request compensation for, and the remodeler shall not be liable for any of the following items:

(1) Incidental, consequential, secondary, or punitive damages
(2) Damages for aggravation, mental anguish, emotional distress, or pain and suffering
(3) Attorney's fees or costs ___________
(owner's initials)

C. The remodeler hereby limits the duration of all implied warranties, including the warranties of workmanship and materials to one (1) year from the date of sale or the date of substantial completion, whichever comes first. ___________
(owner's initials)

D. These limitations shall be enforceable to the extent permitted by law. Some states do not allow the exclusion or limitation of incidental or consequential damages or the limitation of implied warranties, so the limitations or exclusions listed above may not apply.

[Alternative to 3C

[C. The owner understands that no implied warranties whatsoever apply to the structure of the [remodeling, renovation, rehabilitation, or restoration of or addition to] the house and items that are functionally part of the [remodeling, renovation, rehabilitation, or restoration of or addition to] the house. The remodeler disclaims any implied warranties, including (but not limited to) warranties of workmanship and materials to the extent allowed by law, and any implied warranty that exists despite this disclaimer is limited to a period of one (1) year. These limitations shall be enforceable to the extent permitted by the law. Some states do not allow limitations on how long an implied warranty lasts, so this limitation may not apply. ___________
(owner's initials)

Figure 6-3. Remodeler's Sample Limited Warranty Agreement That Excludes Items Covered by the Magnuson-Moss Act (continued)

[The owner acknowledges acceptance of these limitations on the warranties offered by the remodeler in consideration for this limited warranty and the other provisions of the construction contract. Therefore the owner agrees to these limitations if, notwithstanding the provisions of the limited warranty, liability should arise on the part of the remodeler.]

D. Notwithstanding the provisions of this limited warranty agreement, if any liability arises on the part of the remodeler, the remodeler will pay the amount of actual provable damages arising from such liability, but the amount shall not exceed $________. This amount, fixed as liquidated damages and not as a penalty, shall be the remodeler's complete and exclusive amount of liability. The provisions of this paragraph apply if loss or damage results directly or indirectly to persons or property from the meeting or failing to meet the obligations imposed by the construction contract or from negligence, active or otherwise, of the remodeler, the remodeler's agents, or employees.

The owner (a) understands that this provision limits the damages for which the remodeler will be liable and (b) acknowledges acceptance of this liquidated damages provision in consideration for the limited warranties provided by the remodeler and the other provisions of the construction contract. Therefore the owner agrees to this liquidated damages clause if, notwithstanding the provisions of this limited warranty, liability should arise on the part of the remodeler.

____
(owner's initials)

E. This warranty is personal to the original owner of the [remodeling, renovation, rehabilitation, or restoration of or addition to] the house and does not run with the property, the [remodeling, renovation, or rehabilitation of or addition to] the house or with the items contained in the house. The original owner may not assign, transfer, or convey this warranty without the prior written consent of the remodeler.

4. How to Obtain Service

If a problem develops during the warranty period, the owner should notify the remodeler in writing at the address given above of the specific problem. The written statement of the problem should include the owner's name, address, telephone number, and a description of the nature of the problem. The remodeler will begin performing the obligations under this warranty within a reasonable time of the remodeler's receipt of such a request and will diligently pursue these obligations.

Repair work will be done during the remodeler's normal working hours except where delay will cause additional damage. The owner agrees to provide the remodeler or remodeler's representative access to the house and to make available during the work a responsible adult with the authority to approve the repair and sign an acceptance of repair ticket upon completion of the repair.

5. Specific Legal Rights

This limited warranty gives the owner specific legal rights, and the owner may also have other rights that vary from state to state.

6. Where to Get Help

If the owner wants help or information concerning this warranty, the owner should contact the remodeler.

7. The Only Warranty Given by the Remodeler

The owner acknowledges (a) that the owner has thoroughly examined the [remodeling, renovation, rehabilitation, or restoration of or addition to] the house that is to be conveyed, (b) the buyer has read and understands the limited warranty, and (c) the remodeler has made no guarantees, warranties, understandings, or representations (nor have any been made by any representatives of the remodeler) that are not set forth in this document.

I acknowledge having read, understood, and received a copy of this limited warranty agreement.

________ (owner)

________ (remodeler)

Date ________

By ________

Title ________

Date ________

Figure 6-4. Sample Limited Warranty Agreement Conforming to the Magnuson-Moss Act for Builders or Remodelers

> **Warning**—This warranty form must be adapted to conform to applicable state law. This form is only one possible form of warranty conforming to the Magnuson-Moss Act. Builders and remodelers should consult their attorneys for a more complete discussion of a builder's and a remodeler's duties under the act and alternative language. Other language may provide more complete protection for a builder and a remodeler than is included here.

1. Consequential and Incidental Damages

Consequential and incidental damages are excluded, and the implied warranties are limited in duration.

2. Term

The terms of the various coverages of this warranty begin on—

[**For Builders**—The date of final settlement or the date when the buyer first occupies the home, whichever comes first]

[**For Remodelers**—The date of substantial completion (which is the date when the [remodeling, renovation, rehabilitation, or restoration of or addition to] the house renders the house usable for the purpose(s) for which the work was intended)]

3. Coverage

The [builder or remodeler] warrants that by the standards of construction relevant in (city, state) for a period of one (1) year—

A. The floors, ceilings, walls, and other internal structural components of the [home or remodeling, renovation, rehabilitation, or restoration of or addition to the home] that are not covered by other portions of this limited warranty will be free of defects in materials or workmanship.

B. The plumbing, heating, and electric wiring systems, and the septic tank (if the [builder or remodeler] installed it), will be free of defects in materials or workmanship.

C. The roof will be free of leaks caused by defects in materials or workmanship.

The [builder or remodeler] warrants that by the standards of construction relevant in (city, state) for a period of 60 days that the following items will be free of defects in materials or workmanship: doors (including hardware); windows; jalousies; electric switches, receptacles, and fixtures; caulking around exterior openings; plumbing figures; and cabinet work.

4. Manufacturers' Warranties

The [builder or remodeler] assigns and passes through to the [buyer or owner][1] (to the extent they are assignable), the manufacturers' warranties on all appliances and equipment. The following items are examples of such appliances and equipment, although not every [house or remodeling, renovation, rehabilitation, or restoration of or addition to a house] includes all of these items and some [homes or remodelings, renovations, rehabilitations, or restorations of or additions to may include appliances or equipment not in this list: refrigerator, range, furnace or heat pump, washing machine, dishwasher, garbage disposal, ventilating fan, air-conditioner.

5. Exclusions from Coverage

The [builder or remodeler] does not assume responsibility for any of the following, all of which are excluded from the coverage of this limited warranty:

A. Consequential or incidental damages (Some states do not allow the exclusion or limitation of incidental or consequential damages, so the limitation or exclusion may not apply to you.)

B. Defects in appliances and equipment that are covered by manufacturers' warranties (The [builder or remodeler] has assigned these manufacturers' warranties to the [buyer or owner] to the extent they are assignable. If defects appear in these items, the [buyer or owner] should follow the procedures in these warranties)

C. Damage resulting from ordinary wear and tear, abusive use, or lack of proper maintenance of the [house or the remodeling, renovation, rehabilitation, or restoration of, or additions to the house]

D. Defects that result from characteristics common to the materials used, such as (but not limited to) warping and deflection of wood; fading, chalking, and checking of paint from exposure to sunlight; cracks that occurred in the drying and curing of concrete, stucco, plaster, bricks, and masonry;

1. This form is designed for a single buyer or owner. If more than one buyer or owner is involved, the form should be adapted to accommodate the initials and signature of each of the buyers or owners.

Figure 6-4. Sample Limited Warranty Agreement Conforming to the Magnuson-Moss Act for Builders or Remodelers (continued)

drying, shrinking, and cracking of caulking and weather stripping

E. Defects in items installed by the [buyer or owner] or anyone other than the [builder or remodeler] or, if requested by the [builder or remodeler], by the [builder's or remodeler's] subcontractors

F. Work done by the [buyer or owner] or anyone other than the [builder or remodeler] or, if requested by the [builder or remodeler], by the [builder's or remodeler's] subcontractors

G. Loss or injury attributable to the elements

H. Conditions resulting from condensation on, expansion of, or contraction of materials

I. Paint applied over newly plastered interior walls[2]

6. No Other Warranties

This limited warranty is the only express warranty the remodeler gives. Implied warranties, including (but not limited to) warranties of merchantability, fitness for a particular purpose, habitability, and good workmanship are limited to the warranty period (term) set forth above. Some states do not allow limitations on how long an implied warranty lasts, so this limitation may not apply to you. This limited warranty gives you specific legal rights, and you may have other rights that vary from state to state.[3]

7. Claims Procedure

If a defect appears that the [buyer or owner] thinks is covered by this limited warranty, the [buyer or owner] must write a letter describing it to the [builder or remodeler] and send it to the [builder or remodeler] at the [builder's or remodeler's] office address given below:

Customer Service Representative ______________________

Company______________________

Street______________________

City, state, zip ______________________

Emergency phone number ( ) ______________________

The [buyer or owner] must tell the [builder or remodeler] in writing what times during the day that the [buyer or owner] will be at home, so that the [builder or remodeler] can schedule service calls appropriately. If a delay will cause extra damage (for instance, a pipe has burst), the [buyer or owner] should telephone the builder. Only emergency reports will be taken by phone. Failure to notify the [builder or remodeler] of defects covered under this limited warranty or any implied warranties relieves the [builder or remodeler] of all liability for replacement, repair, and all other damages.

8. Repairs

Upon receipt of the [buyer's or owner's] written report of a defect, if the defective item is covered by this warranty, the [builder or remodeler] will repair or replace it at no charge to the [buyer or owner] within sixty (60) days (longer if weather conditions, labor problems, or material shortages cause delays). The work will be done by the [builder or remodeler] or subcontractors chosen by the [builder or remodeler]. The [builder or remodeler] has sole discretion to choose between repair or replacement.[4]

9. Not Transferable

[**For Builders**—This limited warranty is extended to the buyer only if the buyer is the first purchaser of the home. When the first purchaser sells the home or moves out of it, this limited warranty automatically terminates. It is not transferable to subsequent purchasers of the home nor to the first purchaser's tenants.

[**For Remodelers**—This limited warranty is extended to the owner only if the owner continues to own and live in this house after it is [remodeled, renovated, rehabilitated, restored, or added to] by (remodeler's name). When the owner sells the home or moves out of it, this limited warranty automatically terminates. It is not transferable to subsequent purchasers of the home nor to the owner's tenants.

______________________ (buyer or owner)

______________________ (builder or remodeler)

Date ______________________

By ______________________

Title ______________________

Date ______________________

2. Builders and remodelers should consult Section 6C, Statement of Nonwarrantable Conditions, for further exclusions to be included here.

3. This limitation of implied warranties applies to only implied warranties on consumer products. If it is allowed by state law, builders or remodelers can still exclude or disclaim all implied warranties on the house itself; on the remodeling, renovation, rehabilitation, or restoration of or addition to the house; and on any other items not considered consumer products under the act.

4. Although builders and remodelers should have a written policy for dealing with emergencies, they probably do not want to include it in their contracts or warranties. If they do not define emergencies and response time in a brief statement related to matters affecting health and safety, However, buyers will interpret it to suit themselves.

Figure 6-5. Sample Statement of Nonwarrantable Conditions

This statement of conditions that are not subject to the [builder's or remodeler's] warranties explains some of the changes and need for maintenance that may occur in a [new house or a house that is remodeled, renovated, rehabilitated, or restored or added to] over the first year or so of occupancy. A house requires more maintenance and care than most products because it is made of many different components, each with its own special characteristics.

The [buyer or owner] [1] understands that like other products made by humans, a [a house or a house that is remodeled, renovated, rehabilitated, restored, or added to] is not perfect. It will show some minor flaws and unforeseeable defects and may require some adjustments and touching up.

As described in the limited warranty provided to the [buyer or owner] of which this statement of Nonwarrantable Conditions is made a part, the [builder or remodeler] will correct certain defects that arise during defined time periods after construction is completed. Other items that are not covered by the [builder's or remodeler's] warranty may be covered by manufacturers' warranties.

Some conditions, including (but not limited to) those listed in this statement of nonwarrantable conditions, are not covered under the [builder's or remodeler's] warranties. The [buyer or owner] should read these carefully and understand that the [buyer or owner] has not contracted for the builder or remodeler] to correct certain types of problems that may occur in [the buyer's house or the owner's remodeled, renovated, rehabilitated, restored, or added space] . These guidelines will alert the [buyer or owner] to certain types of maintenance (a) that are the responsibility of the [buyer or owner] and (b) that could lead to problems if they are neglected.

The following list outlines some of the conditions that are not warranted by the [builder or remodeler] . The [buyer or owner] should be sure to understand this list. If the [buyer or owner] has any questions, [he or she] should ask the [builder or remodeler] and feel free to consult an attorney before signing the acknowledgment.

[Of the items listed and discussed below, remodelers might want to include only those that pertain to a particular job in the warranty for that job because many of the items would not apply to every job.]

1. Concrete

Concrete foundations, steps, walks, drives, and patios can develop cracks that do not affect the structural integrity of the building. These cracks are caused by characteristics of the concrete itself. No reasonable method of eliminating these cracks exists. This condition does not affect the strength of the building.

2. Masonry and Mortar

Masonry and mortar can develop cracks from shrinkage of either the mortar or the brick. This condition is normal and should not be considered a defect.

3. Wood

Wood will sometimes check or crack, or the fibers will spread apart because of the drying-out process. This condition is most often caused by the heat inside the house or by exposure to the sun on the outside of the house. This condition is considered normal, and the homeowner is responsible for any maintenance or repairs resulting from it.

4. Sheetrock and Drywall

Sheetrock or drywall will sometimes develop nail pops or settlement cracks, which are a normal part of the drying-out process. These items can easily be handled by the homeowner with spackling during normal redecorating. If the homeowner wishes, however, the [builder or remodeler] will send a worker at the end of one (1) year to make the necessary repairs. The [builder's or remodeler's] repairs will not include repainting.

5. Floor Squeaks

After extensive research and writing on the subject, technical experts have concluded that much has been tried but that little can be done about floor squeaks. Generally floor squeaks will appear and disappear over time with changes in the weather and other phenomena.

6. Floors

Floors are not warranted for damage caused by neglect or the incidents of use. Wood, tile, and carpet all require maintenance. Floor casters are recommended to prevent scratching or chipping of wood or tile, and stains should be cleaned from carpets, wood, or tile immediately to prevent discoloration. Carpet has a tendency to loosen in damp weather and will stretch tight again in dryer weather.

1. This form is designed for a single buyer or owner. If more than one buyer or owner is involved, the form should be adapted to accommodate the initials and signature of each of the buyers or owners.

Figure 6-5. Sample Statement of Nonwarrantable Conditions (continued)

7. Caulking

Exterior caulking and interior caulking in bathtubs, shower stalls, and ceramic tile surfaces will crack or bleed somewhat in the months after installation. These conditions are normal and should not be considered a problem. Any maintenance or repairs resulting from them are the homeowner's responsibility.

8. Bricks Discoloration

Bricks may discolor because of the elements, rain run-off, weathering, or bleaching. Efflorescence—the formation of salts on the surface of brick walls—may occur because of the passage of moisture through the wall. Efflorescence is a common occurrence, and the homeowner can clean these areas as the phenomenon occurs.

9. Broken Glass

Any broken glass or mirrors that are not noted by the _[buyer or owner]_ on the final inspection form are the responsibility of the _[buyer or owner]_.

10. Frozen Pipes

The _[buyer or owner]_ must take precautions to prevent freezing of pipes and sillcocks during cold weather, such as removing outside hoses from sillcocks, leaving faucets with a slight drip, and turning off the water system if the house is to be left for extended periods during cold weather.

11. Stained Wood

All items that are stained will normally have a variation of colors because of the different textures of the woods. Because of changes in weather doors that have panels sometimes dry out and leave a small space of bare wood, which the homeowner can easily touch up. These normal conditions should not be considered defects.

12. Paint

Good-quality paint has been used internally and externally on this home. Nevertheless exterior paint can sometimes crack or check. The source of this defect is most often something other than the paint. To avoid problems with the paint, _[buyers or owners]_ should avoid allowing lawn sprinklers to hit painted areas, washing down painted areas, and so on. _[Buyers or owners]_ should also not scrub latex-painted inside walls and should be careful of newly painted walls as they move furniture. The best paint will be stained or chipped if it is not cared for properly. Any defects in painting that are not noted at final inspection are the _[buyer's or owner's]_ responsibility.

13. Cosmetic Items

The _[buyer or owner]_ has not contracted with the _[builder or remodeler]_ to cover ordinary wear and tear or other occurrences subsequent to construction that affect the condition of features in the home. Chips, scratches, or mars in tile, woodwork, walls, porcelain, brick, mirrors, plumbing fixtures, marble and Formica tops, lighting fixtures, kitchen and other appliances, doors, paneling, siding, screens, windows, carpet, vinyl floors, cabinets, and the like that are not recognized and noted by the _[buyer or owner]_ at the final inspection are nonwarrantable conditions, and the upkeep of any cosmetic aspect of the _[house or the remodeled, renovated, rehabilitated, restored, or added space]_ is the _[buyer's or owner's]_ responsibility.

14. Plumbing

Dripping faucets, toilet adjustments, and toilet seats are covered by the _[builder's or remodeler's]_ warranty for a ____________-day (_____) period only. After that, they are the _[buyer's or owner's]_ responsibility. If the plumbing is stopped up during the warranty period and the person servicing the plumbing finds foreign materials in the line, the _[buyer or owner]_ will be billed for the call.

15. Alterations to Grading

The _[buyer's or owner's]_ lot has been graded to ensure proper drainage away from _[the home or the remodeled, renovated, rehabilitated, restored, or added space]_. Should the _[buyer or owner]_ want to change the drainage pattern because of landscaping, installation of patio or service walks, or other reasons, the _[buyer or owner]_ should be sure to retain a proper drainage slope. The _[builder or remodeler]_ assumes no responsibility for the grading or subsequent flooding or stagnant pool formation if the established pattern is altered.

16. Lawn and Shrubs

The _[builder or remodeler]_ accepts no responsibility for the growth of grass or shrubs. Once the _[builder or remodeler]_ grades, seeds and/or sods, and fertilizes the yard, the _[buyer or owner]_ must water the plants and grass sufficiently, and plant ground cover where necessary to prevent erosion. The builder will not regrade a yard, nor remove or replace any shrubs or trees, except for those that are noted as diseased at final inspection.

Figure 6-5. Sample Statement of Nonwarrantable Conditions (continued)

17. Roof

During the first year the warranty on the [buyer's roof or the roof of an owner's addition] is for workmanship and materials. After that the warranty on the roof is for material only and is prorated over the period of the lifetime use of the roof. Warranty claims for any defects in materials will be handled with the manufacturer with the builder's or remodeler's assistance. The builder or remodeler will not be responsible for any damages caused by walking on the roof or by installing a television antenna or other item on the roof.

18. Heating and Air-Conditioning

The [buyer's or owner's] source of heating and air-conditioning is covered by a manufacturer's warranty. The buyer is responsible for making sure the filters are kept clean and changed every thirty (30) days. Failure to do so may void the warranty. Having the equipment serviced or checked at least yearly is a good idea.

19. Indoor Air Quality

[An appropriate disclaimer and warning regarding possible indoor air quality problems, including radon, should be inserted here by the builder or remodeler (see Chapter 5).]

I acknowledge having read and understood and received a copy of the outline above of nonwarrantable items. I understand and agree that these are conditions for which we have not contracted and for which I will not hold the builder liable.

______________________	______________________
(buyer or owner)	(builder or remodeler)
Date ______________	By ______________
	Title ______________
	Date ______________

Seven • **Inspections**

All the contracts in the world will not do as much good for a builder or remodeler as nurturing the relationship with a buyer or owner. One way the builder can nurture the relationship is to have the buyer visit the site often so that he or she can see the difficulties the builder goes through. These visits can be formalized as inspections. This process is somewhat built into a remodeling project when the owner is living on the premises of the home on which the remodeler is working. But even in such cases, remodelers should incorporate formal owner inspections into the remodeling process.

A builder or remodeler can limit many claims by inspecting the property with the buyer or owner and recording the process. Experience has demonstrated that improved communication lessens the chance that a consumer will file a lawsuit. Many problems arise because buyers or owners have unrealistic expectations or simply do not understand the complexities of the construction or remodeling process. The construction or remodeling contract, the limited warranty, the statement of nonwarrantable conditions, and all of the other documents involved in the construction or remodeling process should explain the process to the buyer or owner in simple and understandable terms. These documents also may serve as evidence of what actually occurred during the construction or remodeling process if a dispute should arise.

Interim Inspections

When an owner or the owner's architect or engineer has accepted work after inspecting it, the builder or remodeler may not be liable thereafter for defects in that work. In the absence of an agreement to the contrary, payment for a job may amount to acceptance of the work. Accordingly builders and remodelers should consider having the owner or the owner's agent inspect the work at various stages of completion. The builder or remodeler should record the names of the persons participating in the inspection as well as the scope of the inspection and have the parties sign that record.

Presettlement Buyer Orientation

Home Maintenance Instruction

Before a buyer moves into a new house, the builder or the builder's representative should walk through the house with the buyer and explain what has been done and how the equipment, appliances, and other items work. During this home buyer orientation, the builder should explain to the buyer how to maintain and take care of the items for which the buyer is responsible. Educating the buyer and completing the checklist in Figure 7-1 may prevent problems that result from poor maintenance and that the buyer may try to blame on the builder. At least the buyer will see and will have in writing the items that are his or her responsibility and, thus, may be deterred from trying to blame the builder.[62]

A remodeler should hold the same type of orientation for the owner of a remodeling project and provide similar instruction on maintenance and owner responsibility.

Figure 7-1 lists the items that the builder should discuss with the buyer. It also serves as a record of the builder's disclosures to the buyer. This sample does not attempt to include all of the items that should be addressed by the builder, and every house also may not have every item on the list. The builder should develop and use a format and list that works best for each particular situation.

The remodeler also should adapt this list to suit the owner orientation for a particular project. Should a dispute arise with an owner, the remodeler likewise will benefit from a record of the disclosures made in going over the checklist with an owner and in having the owner sign it.

Figure 7-1. Sample Home Maintenance Instruction Checklist

[Both builders and remodelers would need to adapt this form to their particular homes or projects. For a particular job, a remodeler might use only what was applicable to that job.]

Date ______________________________

Buyer's or owners name ____

Property street address ____

	Buyer's or Owner's Initials
Plumbing	
1. Instruction on use of faucets (emphasize cleaning of aerator); show water cutoffs; and explain how to change washers or cartridges (if used).	_____
2. Instruction on use of shower and tub drains and care of fiberglass surfaces; show water cutoffs.	_____
3. Instruction on adjustment of toilet tank mechanism.	_____
4. Explain care of hot water heater:	_____
A. Periodically check pop-off valve and line.	_____
B. If water is not hot, check the pilot first.	_____
C. Always turn off pilot (or circuit breaker with electric water heaters) before draining tank completely.	_____
D. Periodically flush tank.	_____
5. Show location of water meter, main cutoff, and sink and sewer cleanouts. (In case of an emergency, remove cap and allow sewer to overflow outside.)	_____
6. Show location of any plumbing access panels and their purpose.	_____
7. Show location of well and septic tank (if any), provide instruction on taking care of well pump, tank, and field and explain pumping of tank.	_____
Electrical	
8. Show buyer which outlets are switch-controlled.	_____
9. Explain correct bulb size for lighting fixtures.	_____
10. Show location of main entrance panel.	_____
11. Explain circuit breakers and ground-fault circuit breakers and how to operate them.	_____
12. Explain operation and testing of smoke detectors.	_____
13. Explain operation of security system, intercom, telephone, doorbells, and cable connection for television (if any of these are installed). Show all electrical outlets.	_____
Heating and Air-Conditioning	
14. Explain warranty.	_____
15. Show location and operation of thermostats.	_____
16. Show location of filters and explain when they should be changed.	_____
17. Explain how to balance forced-air systems in heating and cooling seasons.	_____
18. Explain operation of heat pump (if installed) and back-up electric resistance heat.	_____
19. Explain operation of hot water heating system and zones or electric base-board heating (if installed).	_____
20. Explain that, if problems with heating or cooling develop, buyer should do the following in the order listed:	_____
A. Check thermostat setting.	_____
B. Check circuit breaker box to be sure circuit breaker is on (explain that a tripped circuit breaker should be turned all the way off first, then turned back on).	_____
C. Reset circuit breaker.	_____

Figure 7-1. Sample Home Maintenance Instruction Checklist (Continued)

	Buyer's or Owner's Initials
21. Furnish name of company to call for service.	_____
Appliances	
22. Instruction on use and care of all appliances.	_____
23. Explain limited warranties and give buyer other literature.	_____
24. Explain maintenance of range hood and cleaning or changing of range hood filter.	_____
25. Explain how to clean dryer vent pipe and filter.	_____
26. Explain steps to follow if appliances do not operate:	
A. Check to be sure they are plugged in.	_____
B. Check circuit breakers (see item 20).	_____
C. Reset circuit breakers.	_____
27. Furnish names, addresses, and phone numbers of service companies to call for direct service.	_____
General Interior	
28. Explain care, cleaning, and treatment of floors: wood, carpet, tile. Suggest use of casters under furniture. Explain that damage caused by neglect is not warranted.	_____
29. Explain care of paint (not warranted) and that builder does not do touch ups. Remind buyer not to scrub latex-painted interior walls. Give buyer names and color numbers of paints used.	_____
30. Explain use of spackling for normal cracks in sheetrock and nail pops (often repaired by builder after 1 year with owner doing the painting).	_____
31. Explain use of caulk for cracks in tile and to reseal tub when required (not warranted).	_____
32. Explain care of counter tops and that knives will cut the surface.	_____
33. Explain care of doors.	
A. Clean weep holes in sliding glass door thresholds.	_____
B. Replace weatherstripping when it wears out or is damaged.	_____
C. Oil hinges if doors squeak.	_____
D. Avoid paint buildup on door and window sash edges to keep them from sticking.	_____
34. Explain operation of fireplace (if installed), use of seasoned wood to prevent creosote build-up, and the necessity of having flue cleaned regularly.	_____
35. Explain types of cleaning products, such as nonabrasive materials to be used on counter tops, wood finishes, bathroom tile, fiberglass showers and tubs, porcelain, marble and cultured marble, and other products (if used).	_____
36. Explain that bathroom and privacy locks can be unlocked from the outside with small screwdriver, nail, or coat hanger.	_____
37. Demonstrate removal of window sash for cleaning.	_____
38. Explain operation of exterior door and window locks.	_____
39. Explain hairline shrinkage cracks in concrete floors and walks.	_____
40. Explain possible future condensation on cool basement walls and floor if warm, moist air comes in contact with them. Explain avoidance through heating and dehumidification. Warn against trying to dry house out too quickly with excessive heat.	_____
41. Explain how to use bathroom and kitchen exhaust fans to avoid moisture buildup, condensation, and mold.	_____
General Exterior	
42. Show location of and explain secondary air-conditioning drain.	_____

Figure 7-1. Sample Home Maintenance Instruction Checklist (Continued)

	Buyer's or Owner's Initials
43. Explain hairline shrinkage cracks in concrete surfaces.	_____
44. Explain that sunken utility lines and washed-out areas are not warranted.	_____
45. If shrubs are not warranted, explain this fact to the owner. (Be certain that grass and shrubs are alive during inspection.) Explain need for fertilizer and plenty of water.	_____
46. Light fixtures (bulbs not warranted).	_____
47. Explain that hairline cracks in concrete pad and puddles of $1/4$ inch are normal.	_____
48. Explain that treated lumber used in wood decks, steps, or railings should weather 1 year prior to staining.	_____
49. Provide warranties on garage door openers and name, address, and phone number of subcontractor.	_____
50. Wood siding (if used) must be repainted or stained every 4 to 5 years. Mildew can be removed by scrubbing with weak water-and-bleach solution. Aluminum siding can be painted. It will dent if struck. Vinyl siding can crack or break if struck.	_____
51. Explain how to take care of brick.	_____
52. Provide limited warranties on asphalt or other roofing.	_____
53. Explain care of windows and operation of window locks.	_____
54. Provide limited warranties on glass in windows and sliding glass doors.	_____
55. Other ______________________________	_____
______________________________	_____
______________________________	_____
______________________________	_____

[For Builders—I have discussed each of the items listed above with a representative from _(name of builder)_ and I understand them. I have been instructed in the use and care of the above-listed items in my new home and find my home completed in a manner satisfactory and acceptable to me.]

[For Remodelers—I have discussed each of the items listed above with a representative from _(name of remodeler)_ and understand them. I have been instructed in the use and care of the above-listed items in the _[remodeling, renovation, rehabilitation, restoration of or addition to]_ my house. I find this project completed in a manner satisfactory and acceptable to me.]

______________________________ (buyer's or owner's signature)

______________________________ (builder or remodeler)

By ______________________

Title ______________________

Date ______________________

1. This form is designed for a single owner or buyer if more than one owner or buyer is involved, the form should be adapted to accommodate the initials and signature of each of the owners or buyers.

New Home Punchlist

While the builder demonstrates the various items in the house to the buyer or subsequent to those demonstrations, the buyer must inspect the house for defects, problems, or any aspects of construction with which the buyer is dissatisfied. The construction contract should require the buyer to make this inspection before the closing (see Inspection, Acceptance, and Possession in Chapter 2).

As the builder walks through the house with the buyer prior to closing, the buyer fills out the punchlist (Figure 7-2 and 3). The buyer should be required to (a) initial every single item that he or she approves, (b) specifically identify items that require improvement, and (c) spell out what is wrong. These items should be

Figure 7-2. Sample Punchlist Instruction Letter

[The builder presents this letter to the buyer along with the Buyer's Checklist at the time of the pre-settlement buyer orientation to prepare the punchlist. A remodeler would need to develop a letter appropriate to his or her specialty that could be adapted to a particular type of job.]

[Date]

[homebuyer's name]
[street address]
[city, state, and zip code]

Dear [homebuyer's name] :

(builder's name) is proud to welcome you to your new home. Attached to this letter, you will find a checklist to help you inspect your new home. A separate sheet has been provided for each room and area of construction.

Please go through your new home, room by room, and carefully check to see whether all items are in satisfactory condition. Initial the space provided after you have satisfied yourself that the item's condition meets with your approval. If any repairs or adjustments are needed describe the problem in the Improvement Needed column, use the back of the page if necessary. Be sure to write in any additional items that are not listed and describe their condition fully. If a listed item does not apply, put an *X* in the space for your initials.

Before you occupy your home, we will do our best to bring the items that require improvement up to satisfactory condition, consistent with the standards of construction in (city, state), and with our builder's limited warranty. You will be required to make a second inspection at that time, and you will have the opportunity to check all the items to make sure they meet with your approval.

Sincerely,

____
(name of builder or builder's representative)

____
(title)

corrected before closing and before the buyer is allowed to move into the house. Special provisions should be made for corrections that will require more time (for example, if the builder needs to order parts).

The buyer and the builder should agree on a timetable for the improvements and write it on the punchlist. They should also set a date—30 days or so after the closing—for a final inspection, so the builder can check on the minor adjustments that are warranted for a 30-day period (such as dripping faucets and sticking doors).

To avoid the problem of a buyer repeatedly calling a builder to report minor problems in the days or weeks after moving in, the builder might suggest that the buyer keep a record of all minor problems and present it to the builder at one time.

As items on the Improvement Needed part of the checklist are corrected, the buyer should initial and date each item. The builder must keep this punchlist on file until the statute of limitations has expired. If used properly, the punchlist is good evidence of the condition of the house and of the buyer's satisfaction with the condition of the house. Used with the limited warranty, the punchlist provides further limitations on a builder's liability if the warranty is designed to exclude liability for damage or defects that the buyer should have discovered and listed for correction during the inspection.

The list may also serve as evidence that the buyer was satisfied with the items in the house at one time. It also demonstrates the builder's efforts to identify and correct problems.

The punchlist format presented here is merely a suggestion. Different items may apply to other houses, and the builder should consider carefully what format works best in a given situation. The builder could provide a separate punchlist for each room in the house. The list for each room should bear the name of the room to which it applies, and the same format should be used before and after the closing.

Remodeling Punchlist

Although a remodeling project does not involve a settlement, a precompletion punchlist is equally crucial in a remodeling project. The precompletion checklist signed by the homeowner helps to prevent unrelated problems from being blamed on the remodeler. Remodelers could adapt the builders' procedures outlined above and suit them to individual projects.

Figure 7-3. Buyer's or Owner's Checklist for Developing the Punchlist

[The items listed under Improvement Needed will serve as the punchlist. In adapting this sample form, both builders and remodelers should provide more space for writing, especially under Improvement Needed, so the buyer or owner has room to explain what needs to be done. Remodelers probably will want to use only the items that are appropriate for a specific job. For instance, if a kitchen was not involved in a remodeling project, the remodeler would eliminate the kitchen items from the owner's list for that job.]

Date____________________

Item	Initials	Improvement Needed
Bathroom		
Vanity	____	__________
Sink	____	__________
Medicine cabinet	____	__________
Bathtub	____	__________
Shower	____	__________
Shower curtain bar or door	____	__________
Toilet	____	__________
Towel bars	____	__________
Paper holder	____	__________
Floor	____	__________
Walls	____	__________
Ceiling	____	__________
Light fixture (not bulbs)	____	__________
Windows	____	__________
Doors	____	__________
Woodwork	____	__________
Baseboard heater or air vent	____	__________
Other	____	__________
Kitchen		
Sink	____	__________
Cooktop	____	__________
Oven, range	____	__________
Hood and exhaust fan	____	__________
Microwave	____	__________
Mixer	____	__________
Dishwasher	____	__________
Refrigerator	____	__________
Freezer	____	__________
Disposal	____	__________
Trash compactor	____	__________

Item	Location	Initials	Improvement Needed
Cabinets		____	__________
Drawers		____	__________
Countertops		____	__________
Floor		____	__________
Walls		____	__________
Ceiling		____	__________
Light fixture (not bulbs)		____	__________
Windows		____	__________
Doors		____	__________
Woodwork		____	__________
Baseboard heater or air vent		____	__________
Other		____	__________
Living Room, Dining Room, Den, Family Room, Bedroom[1]			
Floor	____	____	__________
Walls	____	____	__________
Ceiling	____	____	__________
Ceiling fan	____	____	__________
Light fixtures (not bulbs)	____	____	__________
Doors	____	____	__________
Windows	____	____	__________
Shades	____	____	__________
Woodwork	____	____	__________
Baseboard heater or air vent	____	____	__________
Closets	____	____	__________
Closet rods and shelving	____	____	__________
Other	____	____	__________

Item	Location	Initials	Improvement Needed
General Interior			
Interior doors	____	____	__________
Hardware	____	____	__________
Paneling	____	____	__________
Insulation	____	____	__________
Wallpaper, paint	____	____	__________
Fireplace(s)	____	____	__________
Cabinets	____	____	__________
Bookcases	____	____	__________
Woodwork	____	____	__________
Tile work	____	____	__________
Electrical switches	____	____	__________
Electrical outlets	____	____	__________
Glass	____	____	__________
Hallway[1]			
Walls	____	____	__________
Floor	____	____	__________
Ceiling	____	____	__________
Light fixture not bulbs)	____	____	__________
Doors	____	____	__________
Closets	____	____	__________
Windows	____	____	__________
Shades	____	____	__________
Woodwork	____	____	__________
Basement or Utility Room			
Washer	____	____	__________
Dryer	____	____	__________

Item	Location	Initials	Improvement Needed
Water heater	____	____	__________
Furnace or heat pump	____	____	__________
Closets	____	____	__________
Walls	____	____	__________
Floor	____	____	__________
Ceiling	____	____	__________
Light fixture (not bulbs)	____	____	__________
Light switches	____	____	__________
Doors	____	____	__________
Baseboard heater or air vent	____	____	__________
Other	____	____	__________
General Exterior			
Paint		____	__________
Siding, brick		____	__________
Chimney		____	__________
Roof		____	__________
Doors		____	__________
Garage		____	__________
Walkways		____	__________
Balcony		____	__________
Light fixtures (not bulbs)		____	__________
Light switches		____	__________
Electrical outlets		____	__________
Vents (dryer, range, hood, bathroom fan)		____	__________
Other		____	__________

I understand each of the items on the preceding pages and have discussed each item with a representative of (builder's name). I have been instructed in the use and care of the items listed on the preceding pages of this punchlist. With the sole exception of the items listed under Improvement Needed, I find my [new home or project] to be completed in a manner satisfactory and acceptable to me. I understand that, with the exception of those items that I have listed under Improvement Needed, I am purchasing the [house or project] as is, and I understand that the [builder or remodeler] makes no other guarantees or warranties other than those that are clearly stated in the contract and the other contract documents.

Punchlist completion date ____________________

Reinspection Date ____________________

____
(buyer's or owner's signature)

____
(builder or remodeler)

Date ____________________

By ____________________
(signature)

Title ____________________

Eight • Subcontract

Building and remodeling companies rarely perform all of the construction work on a project using their own employees as a labor source. Most of the labor and much of the material is provided by subcontractors—people or companies that contract with builders and remodelers to provide specialized services and materials. Sometimes the term *subcontractor* is defined for a specific purpose, such as when it used in a mechanic's lien statute, but for our purposes a subcontractor has a direct contractual relationship with a builder or remodeler to undertake a specific part of the work to fulfill a builder's or remodeler's contract with a buyer or owner.

Builders and remodelers should have written contracts with their subcontractors for the same reasons that they have written contracts with their clients: to document their agreement, to minimize misunderstandings, and to allocate risks. The complexity of the subcontract may depend in part on the builder's or remodeler's relationship with the subcontractor, and on the subcontractor's level of business sophistication. In situations in which the parties have a longstanding business relationship or in which subcontractors are accustomed to making their proposals on scrap paper, the parties may want to start out with a short subcontract limited to the key provisions such as scope of the work, contract documents, contract price, payment schedule, time for performance, insurance, changes, warranty, clean-up, safety, and dispute settlement (arbitration and/or mediation). When the parties are comfortable using the written contract, they can add more detail and add additional provisions such as liquidated damages, indemnification, lien waivers, rules and regulations regarding conduct while on the site, and so on.

A written subcontract also may help establish the subcontractor's status as an independent contractor, rather than as an employee. The determination of whether an individual is an employee or an independent contractor can have a wide range of ramifications for an employer. For example, certain labor laws, such as the Fair Labor Standards Act, which provides minimum wage and overtime standards, covers employees only. Similarly a number of state labor laws, workers' compensation laws, and unemployment insurance laws apply only to employees. An employer is liable for the wrongful acts his or her employee commits while on the job. On the other hand, with certain exceptions, an employer is not liable for physical harm caused by an act or omission of the independent contractor or his or her employees or agents.

The distinction between an independent contractor and an employee is also relevant if a worker is injured. Generally the exclusive remedy of an employee injured on the job is workers' compensation, while an independent contractor injured on the job may sue the builder or remodeler for his or her injuries.

Federal and state tax withholding obligations differ depending on the classification of a worker. Failure to properly classify an person as an employee may result in back taxes and penalties.

The provisions discussed below do not address every contingency, nor do they apply to all contracts with subcontractors. Thus builders or remodelers must carefully consider the facts of each situation and use a subcontract specifically designed to fit those facts. Because subcontracts are legal documents that have great impact on builders' or remodelers' liability, builders and remodelers should have their attorneys prepare or, at least, review such documents before signing them.

Subcontractor's Proposal

The proposal is usually the blueprint for the offer to enter into a subcontract. The proposal's description of the scope of the work should be precise. The description should state the plans and specifications by date, date of last revisions, and addenda.

When the builder or remodeler signs a subcontractor's proposal it becomes a binding subcontract. A builder or remodeler also may incorporate the terms and conditions of the proposal into any contract that is subsequently signed by adding the words: "The scope of the work and the terms and conditions as stated in (subcontractor's) proposal (dated the ________ day of __________, 19____) are incorporated by reference and shall take precedence over any conflicting provisions in the contract." Of course, the language of this or any other subcontract provision can be changed according to the wishes and bargaining strength of the various parties (see Figure 8-1).

The Subcontract

Scope of the Work

The clause describing the work to be performed in a subcontract between a builder or remodeler and a subcontractor must be precise. Otherwise the builder or remodeler may expect work that the subcontractor did not contemplate doing when bidding on the job or negotiating the subcontract, or the subcontractor may not perform work that the builder or remodeler expected to be done. As previously discussed throughout this book, unmet expectations often cause disputes between the owner or buyer and the builder or remodeler. Similarly unmet expectations can cause disputes between the builder or remodeler and the subcontractor and result in litigation.

Accordingly the description of the work should contain at least three basic elements: (a) a detailed description of the work to be done and, importantly, work that will not be done by the subcontractor, (b) a designation of the relevant plans by date and date of last revision, and (c) a statement that the work shown on the

Figure 8-1. Sample Proposal

For:

____
(name of builder or remodeler)

____
(street address)

____
(city, state, zip)

Phone (____)______________________

Drawings No.____ Dated ______________

Date ______________________

Project ______________________

Location ______________________

____

Design Professional ______________

Specifications Dated ______________

Subject to prompt acceptance within ______________ (____) days and to all conditions stipulated below, we propose to furnish materials and labor at the price(s) stipulated below:

Price: $__________

The undersigned accepts this proposal and all its terms and conditions as a binding contract.

____
(builder or remodeler)

By ______________________
(signature)

Title ______________________

Date ______________________

____
(subcontractor)

By ______________________
(signature)

Title ______________________

Date ______________________

designated plans is to be done in accordance with the project specifications.[63] This section should specify whether the subcontractor is expected to provide all of the materials and supplies necessary to perform the subcontracted work. The contract also should list those instances in which the subcontractor is providing only labor; alternatively the contract should describe the materials which the subcontractor will provide.

Price and Payment

To avoid any misunderstandings, the parties should clearly define the payment terms in the subcontract. This section should state the total contract price in words and numbers, and it should describe the method of payment with particular emphasis on when final payment is due. For example, if the subcontractor will be paid in draws, the contract should include a draw schedule showing the number of payments, the timing of each payment, and the administrative preconditions of each payment. Similarly, if the subcontractor will be paid monthly, the contract should clearly define the billing procedure. The contract should specify whether final payment is due upon substantial completion or total completion, or some other milestone; state whether the subcontractor must provide releases and lien waivers; and include any other mutually agreed upon terms.

If the owner's contract with the builder or remodeler allows the owner to withhold retainage, the builder or remodeler also may want to withhold retainage from the subcontractor to assure the subcontractor's completion of the project and correction of defects. The contract should provide for the retainage of a certain percentage from the funds owed to the subcontractor and should state whether the retainage will be paid to the subcontractor at final payment or at some other time. If the retained funds are withheld until substantial completion of the entire project, retainage may be particularly unfair to those subcontractors who participated in the early stages of the project. For those subcontractors, the builder or remodeler should tie the release of retainage to substantial completion of the subcontractor's work rather than to substantial completion of the entire project.

> The [builder or remodeler] agrees to pay the subcontractor for the performance of this subcontract the sum of ____________________________ Dollars ($_______________), subject to additions and deductions for such changes as may be agreed upon in writing. Partial payments will be made to the subcontractor as follows:__________________.
>
> Each request for payment shall list the amount paid for any labor not covered by worker's compensation insurance. The premium for worker's compensation insurance for such uninsured labor shall be withheld from the amount due.

Change Orders

The change order clause in a subcontract offers protection to the builder or remodeler because it requires the subcontractor to perform requested changes in the work. The builder or remodeler must retain the right to order changes even when terms have not been agreed upon. The builder or remodeler needs a mechanism to establish a dollar ceiling and other limits for job changes (usually builders' and remodelers' subcontracts require the subcontractors to perform the extra work on a lump-sum amount to be negotiated.

Before allowing the subcontractor to begin any new work created by an owner's request for change, the builder or remodeler should prepare a change order and have the owner sign it. That written change order agreement should specify the revisions in the price, the work, the payment schedule, if necessary, and the completion date, if necessary.

The subcontract should clearly identify who is authorized to approve a change order. Usually change orders should come from the person who executed the contract. If some person other than the builder or the remodeler has the authority to approve a change order, the builder or remodeler and the subcontractor should make sure that authority is in writing.

> The [builder or remodeler] may order additional work and the subcontractor will perform these changes in the work. However no alteration, addition, omission, or change shall be made in the work nor in the method or manner of the performance, except upon the written change order of the builder or remodeler. Any change or adjustment in the subcontract price by virtue of the change order shall be specifically stated in the change order. If the [builder or remodeler] and the subcontractor cannot agree upon a fixed price, the [builder or remodeler] may direct the subcontractor to perform the work, and the subcontractor will be paid based upon the actual cost to the

> subcontractor for all materials, labor, equipment, and supervision plus a reasonable overhead and profit [alternate language: plus ____ percent] . No change order shall vary, abrogate, void, or otherwise affect the terms, conditions, and provisions of this subcontract except as specifically stated in the change order.

Indemnification

One method of minimizing liability available to builders and remodelers is allocating the risk of loss to the party directly responsible for the loss. An indemnification clause in the subcontract can accomplish this allocation and help to protect the builder or remodeler from actions brought by the buyer, the owner, or third parties relating to the subcontractor's work. For example, if the builder or remodeler is liable to the homeowner for defective construction by a subcontractor, an indemnification clause in the subcontract will entitle the builder to recover the cost of correcting the work from the subcontractor.

Although builders and remodelers use several types of indemnification provisions, generally the indemnification clause provides that the subcontractor will compensate or defend the builder or remodeler against claims, loss, damage, or expense arising out of or resulting from the subcontractor's work.

> The [builder or remodeler] shall not be liable for any loss or casualty incurred or caused by the subcontractor's work. The subcontractor shall hold the [builder or remodeler] harmless from any and all liability, costs, damages, attorney's fees, and expenses from any claims or causes of action arising while on or near the project, or while performing contract-related work, including those claims relating to the subcontractor's subcontractors, suppliers, or employees, or by reason of any claim or dispute of any person or entity for damages from any cause directly or indirectly relating to any action or failure to act by the subcontractor, its representatives, employees, subcontractors, or suppliers. The [builder or remodeler] may retain any and all monies due to be paid or to become due to be paid to the subcontractor under this or any other contract, sufficient to save itself harmless and indemnify itself against any liability or damage, including attorney's fees.

Insurance

Like the builder and the remodeler, the subcontractor should obtain liability insurance, including automobile and property damage insurance, and where workman's compensation insurance is required by law, the subcontractor must obtain it. If the subcontractor agrees to indemnify or defend the builder or remodeler, the subcontractor's insurance should contain a contractual liability endorsement covering indemnity and defense obligations of the subcontractor.

The contract should specify that, before construction, the subcontractor shall furnish to the builder or remodeler original certificates or copies of policies showing insurance coverage of the desired types and in appropriate amounts. Similarly if the subcontractor is exempt from workers' compensation requirements, the subcontractor should be required to offer proof of exemption prior to construction.

A wise builder or remodeler will not allow the subcontractor to begin work until he or she provides such information. However in actual practice a subcontractor may proceed with the work but fail to provide the required insurance. For these occasions, the contract should provide that, if the subcontractor fails to furnish or maintain the required insurance, the builder or remodeler shall have the right to (a) obtain such insurance for the subcontractor and (b) withhold from the subcontractor's payments the cost of the coverage at the applicable rate for the trade involved or any costs incurred by the builder or remodeler resulting from the subcontractor's failure to furnish or maintain such insurance. The subcontract should include a clause stating: "All certificates of insurance and all insurance policies shall provide that the insurance may not be canceled, terminated, or modified without (________) days advance written notice to the builder or remodeler."

Taxes, Charges, and Permits

This clause assigns responsibility for the taxes, fees, and other charges generated by the subcontractor's work under the subcontract. In addition, in the example below, the subcontractor must secure the necessary permits for completing the subcontractor's portion of the project.

The subcontractor understands and agrees that he or she is an independent contractor and that he or she shall be responsible for and pay any and all taxes, contributions, fees, and similar expenses imposed directly or indirectly for his or her work, labor, material, and services required by or relating to this contract. The subcontractor shall at the subcontractor's own expense apply for and obtain all necessary permits and conform strictly to the laws, ordinances, and regulations applicable in the locality in which the work is performed. At no time shall the contract price increase or escalate on account of any such charge. On demand, the subcontractor shall substantiate that all taxes and other charges are being properly paid.

Clean-Up

The contract should assign responsibility for removal of the debris generated by a subcontractor's work on the job. Clean-up is largely a matter of local custom and practice. In some areas subcontractors drop their trash wherever it is created, and the builder or remodeler cleans it up. In other areas subcontractors customarily disposes of all their own trash in a designated place on the jobsite. In some isolated situations, custom calls for subcontractors to remove all of the debris generated by their work at the job site. This last situation occurs more commonly in multifamily construction. When the builder or remodeler cleans up the subcontractor's trash, that expense is allocated to subcontractor costs.

The subcontractor shall at all times keep the building and premises broom clean of dirt, debris, rubbish, and any other waste materials arising from the performance of this subcontract. The subcontractor is responsible for removal of all debris created by his or her work and the disposal thereof at a designated spot at the jobsite.

Conduit Clause

The conduit clause, also known as the flow-down clause, allows the builder or remodeler to shift risks downward in the contractual chain. The clause binds the subcontractor to the builder or remodeler as the builder or remodeler is bound to the buyer or owner. This provision should state that the builder or remodeler has the same rights and privileges against the subcontractor as the construction contract gives to the buyer or owner against the builder or remodeler.

This type of clause is particularly useful where the prime contract is dictated by the owner, and the builder or remodeler has little input into the terms of the contract. If a dispute between the buyer or owner and the builder or remodeler results in arbitration, this clause may enable the builder or remodeler to join the subcontractor in the arbitration with the owner and avoid a separate action against the subcontractor.

Of course, if the builder or remodeler attempts to bind the subcontractor to the provisions of the general contract, he or she should be prepared for a request from the subcontractor for rights, remedies, and redress corresponding to those the builder or remodeler has against the owner. If a conduit clause is used in the subcontract, the subcontractor is entitled to a copy of the agreement between the buyer and the builder or the owner and the remodeler before signing any agreement with the builder or remodeler. The subcontractor also should inspect all plans and specifications and all contract documents.

The subcontractor shall assume toward the [builder or remodeler] all the obligations and responsibilities that the [builder or remodeler] assumes toward the [buyer or owner] under the general contract, and the [builder or remodeler] shall have the same rights and privileges against the subcontractor as the [buyer or owner] in the general contract has against the [builder or remodeler] insofar as these obligations, responsibilities, rights, and privileges pertain to the subcontractor's work.

Illustrative Cases

Case 1—Perry Avenue Fund 88, Ltd., hired J. R. Slaught Construction Company to supervise the construction of 58 single-family dwellings on property owned by Perry in California. The parties' contract provided for binding arbitration of all disputes. Thereafter Slaught hired several subcontractors and suppliers. A dispute arose between Slaught and the owner, and the matter was submitted to arbitration. In response to Slaught's demand for arbitration, the owner counterclaimed that the work was improperly performed or contained defective material. Slaught demanded that the subcontractors join in the arbitration between Slaught and the owner, but they declined. The California Court of Appeal relied on the following clause in the subcontracts to compel the subcontractors to participate in the arbitration:

> 5. ASSUMPTION OF PRINCIPAL CONTRACT: The work to be done hereunder is a portion of the work required of Contractor under the General Contract referred to in the Special Conditions hereof. Insofar as applicable, Subcontractor shall be bound by all of the terms and conditions of the Contract Documents, and shall strictly comply therewith. All rights and remedies reserved to Owner under the Contract Documents shall apply to and be possessed by Contractor in its dealings with Subcontractor.

The court concluded that, because the Construction Contract required Slaught to submit to arbitration, the arbitration provision as incorporated into the subcontracts through paragraph 5 also required the subcontractors to submit to arbitration.[64]

Case 2—The Alaska State Housing Authority awarded a contract to Wick Construction Company to build a courthouse and office building. Wick subcontracted all work relating to the fabrication and erection of the curtainwall to Kenai Glass Company. The building was completed and accepted 414 days late. Wick attributed the delay to Kenai, and it withheld the final payment due Kenai under the subcontract. Kenai sued for the retainage and Wick counterclaimed that Kenai breached the subcontract by failing to perform in a timely manner. The trial court awarded Wick $765,654 in total damages, and Kenai appealed on the grounds that the subcontract limited Wick's recovery for its damages to $400 per day of delay. The Alaska Supreme Court, finding for Kenai, held that a conduit clause in the subcontract incorporated a liquidated damages clause in the contract between the Housing Authority and Wick, limiting liability for delay to $400 per day.[65]

Concealed Conditions

Just as the prime contract addresses the possibility of changed or differing conditions, the contract between the builder or remodeler and the subcontractor should also address this issue (see Differing Site Conditions in Chapters 2 and 3). Generally, if the subcontract contains no differing site conditions clause and it imposes a site inspection requirement on the subcontractor, the subcontract places the risk of uncertainty of subsurface conditions on the subcontractor. Thus the subcontract might expressly provide that the subcontractor has inspected the site and that in arriving at the contract price the subcontractor has assumed the risk that unforeseen conditions or events may make the job more expensive.

Termination of Agreement

Termination is a remedy for a material breach of contract. The subcontract should spell out the circumstances under which the subcontract may be terminated. Additionally the subcontract should clearly state the rights, duties, and obligations of the parties in the event of termination. Conversely the subcontract may provide for cancellation by the builder or remodeler without cause. Subcontracts sometimes include language that allows the builder or remodeler the unrestricted right to cancel the subcontract agreement. Such a clause is likely to be used by a builder or remodeler whose contract with the owner allows the owner to terminate the contract at any time.

> **For Cause**—In the event the subcontractor should at any time fail to perform the work with promptness or diligence, the [builder or remodeler] shall have the right to terminate this agreement for cause after three (3) days notice to the subcontractor (unless within said three [3] day period the subcontractor begins to remedy such failure). In the event of termination by the [builder or remodeler], the [builder or remodeler] shall finish the subcontractor's work or have it finished. After all payments are made for finishing the work, the builder would pay the subcontractor any balance remaining.
>
> **Without Cause**—The [builder or remodeler] has the option to terminate this agreement without cause. In the event of termination by the [builder or remodeler], the [builder or remodeler] shall pay the subcontractor for work already completed and for loss of anticipated profits.

No Lien

The lien waiver clause typically provides that the builder, remodeler, or subcontractor will not file any liens against the property on account of labor, material, or equipment furnished under the contract. A lien waiver clause offers little, if any, benefit to a builder, remodeler, or subcontractor. A blanket provision in the contract documents waiving any lien rights whatsoever is to be avoided. Lien waivers or releases should relate only to payments received.

As discussed in Chapters 2 and 3, some buyers or owners may want builders or remodelers to waive their lien rights at the beginning of the job. Homeowners find such a provision attractive because it ensures that the property will remain unencumbered by liens. In rare instances, builders or remodelers may waive their lien rights. If they do, they may also require their subcontractors or suppliers to waive their lien rights.

Some states permit builders or remodelers to waive not only their own lien rights but also those of the subcontractor. In other states a lien waiver signed at the beginning of the job and before the work has been performed is unenforceable (see the sample lien waiver statute in Figure 8-2).

If the subcontract contains an incorporation-by-reference clause, the builder or remodeler should inform the subcontractor of any blanket lien waiver in the prime contract and incorporate it into the subcontract by reference. Generally, if a lien waiver is permitted, it must be conspicuous.

Illustrative Case

The Village of Westmont awarded a contract to CBI Na-Con to construct a ground storage reservoir. CBI Na-Con subcontracted the electrical work to Thompson, who contracted with Process to obtain certain equipment required by the specifications. Process shipped the equipment to Thompson in May 1989, but it had not received payment as of September 1989. Thompson offered to have 95 percent of the amount ready within 24 hour notice upon Thompson's receipt of Process's "partial waivers." Process responded that the waivers of liens were not a condition of sale and subsequently sued Thompson for the funds.

Finding for Thompson, the court held that Process proposed to furnish the materials called for in sections 91 and 93 of the general contract specifications and that both sections referred to the general conditions of the general contract. The general conditions provided, in part, that the contractor would procure from each subcontractor and supplier of material or labor a waiver of any claim that they had under the mechanic's lien laws of the state in which the work was located.

The court observed that the specific adoption by the subcontract of the drawings and specifications of the prime contract made those provisions as much a part of the subcontract as if they were expressly written in it, and "Accordingly Process had the duty to learn of or know the terms of the general contract before it contracted to furnish the equipment called for in the general contract specifications."[66]

Safety

The subcontract should provide that the subcontractor will cooperate with the builder or remodeler to prevent injuries to any workers on the job site. In addition the subcontract should confirm the subcontractor's obligation to observe and comply with applicable federal, state, and local safety and health rules and regulations, including the Occupational Safety and Health Act.

Figure 8-2. Pennsylvania Lien Waiver Statute

A written contract between the owner and contractor or a separate written instrument signed by the contractor, which provides that no claim shall be filed by anyone, shall be binding; but the only admissible evidence thereof, . . . against a subcontractor, shall be proof of actual notice thereof to him before any labor or materials were furnished by him; or proof that such contract or separate written instrument was filed in the office of the prothonotary [clerk] prior to the commencement of the work upon the ground or within ten (10) days after the execution of the principal contract or not less than ten (10) days prior to the contract with the claimant subcontractor, indexed in the name of the contractor as defendant and the owner as the plaintiff and also in the name of the contractor as plaintiff and the owner as defendant. The only admissible evidence that such a provision has, notwithstanding its filing, been waived in favor of any subcontractor shall be a written agreement to that effect signed by all those who, under the contract, have an adverse interest to the subcontractor's allegation. Pa. Stat. Ann. tit. 49, §1402 (1903 and Supp. 1995).

In the course of the work the subcontractor shall initiate, maintain, and supervise all safety precautions and programs against injury to persons and property. The subcontractor shall provide safe working conditions for its employees, other employees, and other persons and entities on the site.

In furtherance thereof, the subcontractor shall give all notices and comply with all applicable federal, state, and local laws bearing on the safety of persons or property or their protection from damage, injury, or loss on or about the premises where the work is being performed.

Establishment of a safety program by the [builder or remodeler] shall not relieve the subcontractor or other parties of their safety responsibilities. The subcontractor shall indemnify the [builder or remodeler] for fines or penalties imposed upon the [builder or remodeler] to the extent caused by the subcontractor's failure to comply with applicable safety requirements, and for attorneys fees and costs incurred in defending any citations for noncompliance by the subcontract.

Warranty

The warranty is another risk-allocation device, similar to the indemnification clause, that allows the builder or remodeler to minimize his or her liability. Prudent builders and remodelers will obtain warranties from their subcontractors to cover the quality of the subcontract work and the materials furnished for the period of time required of the builder or remodeler by the owner or purchaser (or some other period of time agreed to by the parties). For example, the warranty might commence with the date the certificate of occupancy is issued, and it might run in accordance with the dates outlined in the builder's or remodeler's limited warranty to the purchaser or owner. Subcontractors will want the warranty to start from the time of completion of their installation rather than from the time of acceptance of the home by the owner or purchaser. The warranty should explicitly state what is covered and what is not covered, and it should conform to certain standards.

The subcontractor warrants his or her work under this subcontract against all deficiencies and defects in materials and/or workmanship. Unless otherwise specified in this subcontract, all materials and equipment furnished shall be new. Substitutions not properly approved or authorized and unauthorized deviations from plans, specifications, manufacturer's installation instructions or building codes shall be deemed to be defects in workmanship covered by this warranty, whether or not damages result. The subcontractor agrees to repair or replace at [his or her] own expense and pay any damages resulting from any defect in materials or workmanship that appear within one (1) year from [the date of occupancy by the owner, completion of the subcontract work, or acceptance or use by the (builder or remodeler)]. Warranty work must be completed promptly [within ___ working days] after receipt of a written request from the [builder or remodeler] or the [buyer or owner]. If the subcontractor does not promptly complete the warranty work, the [builder or remodeler] or the [buyer or owner] may have the defective work corrected, and all direct and indirect costs of such work, including compensation for additional professional services will be paid by the subcontractor. In case of an emergency, as defined in the [builder's or remodeler's] limited warranty the subcontractor shall respond within ___ hours after receiving notice of the emergency. Emergency service may be requested by telephone with a written service order to follow.

Independent Contractor Status

As discussed at the beginning of this chapter a written subcontract may help establish the subcontractor's status as an independent contractor, rather than as an employee. However, actions speak louder than words, and the courts and the government agencies may disregard the contract if the parties treat each other as if they had an employer-employee relationship.

The employment and its affect on insurance and taxes is of particular concern to builders and remodelers. Builders and remodelers who cannot prove that their workers are independent subcontractors, rather than employees, may end up paying insurance premiums as if the subcontractors were employees. During its annual audit, the insurance company will request certificates of insurance for subcontractors, showing the dates and amounts of the subcontractor's own insurance. Because the premiums are calculated as a percentage of payroll, if these certificates are not produced, the insurance company will add the labor costs paid to subcontractors to the builder's or remodeler's payroll and increase the premium accordingly. Therefore builders and remodelers should require that their subcontractors provide them with valid certificates of insurance prior to construction.

Federal tax withholding obligations differ depending on the classification of a worker. Companies hiring employees generally must withhold unemployment tax, Social Security tax, and income tax. In addition employers may have state unemployment and income tax withholding obligations similar to those imposed under federal law. The failure to "properly" classify an independent contractor as an employee may result in back taxes and penalties.

The Internal Revenue Service (IRS) has identified 20 factors it considers in identifying whether an individual is an employee or an independent contractor. No one factor determines the employment relationship. The worker's employment status depends on the total situation. However the degree of control exercised by, or granted to, the employer is generally conceded to be a significant factor. The subcontract should incorporate as many of the 20 factors as is feasible (see Figure 8-3).

If the IRS concludes that a worker is an employee, all is not lost. Congress has created a statutory "safe harbor," which overrides the common law determination (the 20 IRS factors). If certain conditions are met, the Revenue Act of 1978 530(a), 26 U.S.C. 3401 note (1989) allows employers to treat workers as independent contractors, even though under common law they might be considered employees. In order to take advantage of the "safe harbor" provision, the employer must have (a) consistently treated the worker (and any individual holding a substantially similar position) as an independent contractor and (b) had a reasonable basis for not treating the worker as an employee.

Attorney's Fees

The general rule in this country is that each party to a lawsuit pays its own attorney's fees—win or lose. Thus, the prevailing party in a lawsuit often ends-up paying a considerable amount of the judgment to his or her attorney or pays a considerable sum of his or her own money to the attorney for successfully defending a suit (see Chapters 2 and 3 on collecting attorney's fees).

Figure 8-3. IRS's 20 Factors Affecting Independent Contractor Status

1. Instructions—A worker who is required to comply with other persons' instructions about when, where, and how he or she is to work is ordinarily an employee. The right to control, not actual control, is all that is required to establish control.
2. Training—Training an employee is indicative of an employer-employee relationship.
3. Integration—Integration of the worker's services into the business operations generally shows that the worker is subject to direction and control.
4. Services Rendered Personally—Indicative of an employer-employee relationship because presumably the person for whom the services are performed is interested in the methods used to accomplish the work as well as in the results.
5. Hiring, Supervision, and Paying Assistants—An independent contractor should have the right to hire and fire assistants.
6. Continuing Relationship—A continuing relationship between the worker and the employer is indicative of an employer-employee relationship.
7. Set Hours of Work—The establishment of set hours of work by the person for whom the services are performed is a factor indicating control.
8. Full-Time Required—An independent contractor is free to work when and for whom he or she chooses.
9. Doing Work on Employer's Premises—While this factor depends on the nature of the service involved, control over the place of work is indicated when the [person for whom the services are performed has the right to compel the worker to work at specific places as required.
10. Order of Sequence Set—An independent contractor is free to follow his own work pattern and need not follow the established routines and schedules of the person for whom the services are performed.
11. Oral or Written Reports—An independent contractor should not be required to submit regular oral or written reports to the person for whom the services are performed.
12. Payment by the Hour, Week, Month—Payment made by the job or on a straight commission generally indicates that the worker is an independent contractor.
13. Payment of Business and/or Traveling Expenses—An employer generally pays the employee's business and/or traveling expenses.
14. Furnishing of Tools and Materials—An independent contractor furnishes his or her own tools and materials.
15. Significant Investment—Independent contractors invest in facilities that they use in performing services.
16. Realization of Profit and Loss—Independent contractors can realize a profit or suffer a loss as a result of the work they perform.
17. Working for More Than One Firm At A Time—An independent contractor performs more than de minim services for a multiple of unrelated persons or firms at the same time.
18. Making Service Available to the General Public—An independent contract makes his or her services available to the general public.
19. Right to Discharge—An independent contractor cannot be fired so long as he or she meets his or her obligations under the contract.
20. Right to Terminate—Generally, an employee may terminate the employment relationship without incurring liability.

Rev. Ruling 87-41 (1987).

Nine • Contracts with Other Team Members

During the course of a construction or remodeling project, the builder or remodeler may work with other professionals such as attorneys, in-house real estate salespeople, real estate brokers, suppliers, and lenders, to name a few (Figure 9-1). This chapter highlights some of the legal issues that may arise when working with these individuals and entities.

Attorneys

This book advises builders and remodelers throughout that they should not use the forms and procedures in it without the approval of an attorney, preferably one experienced in construction contract law. The list in Figure 9-2 explores some of the ways in which attorneys can assist builders and remodelers with their contracts.

Of course, the role of the attorney is not limited to contract review or drafting. A builder or remodeler may call upon an attorney to provide a number of other services. For example, the attorney may—

- advise the builder or remodeler regarding compliance with federal, state, and local regulations
- assist in obtaining a building permit
- provide support during presentations to planning and zoning boards
- represent the builder or remodeler in financial transactions
- represent the builder or remodeler in a situation in which an owner or subcontractor breaches a contract (such as when an owner refuses to pay or a subcontractor refuses to perform)
- defend a builder or remodeler in the event of a breach of contract or negligence action
- draft contracts
- assist the builder or remodeler in filing a mechanic's lien
- represent the builder or remodeler in dispute resolution proceedings

When hiring an attorney, builders and remodelers should consider the issues discussed in the following paragraphs:

- Does he or she charge hourly, by the case, or on a contingency basis? A contingency arrangement is usually available only when the builder or the remodeler sues someone. Contingency means that the attorney will collect fees only if the builder or remodeler receives money upon winning the case. In such a case, however, the builder or remodeler still usually has to pay expenses.
- If the charge is hourly, what is the rate, and what is the rate for other attorneys in the office who may work on the case?
- Is the rate subject to change?
- In addition to the attorney's fees, is the client responsible for other costs, such as expenses?
- When are fees and costs payable?
- Does the attorney offer a mechanism for resolving client disputes?
- Under what circumstances may the parties terminate the relationship, and how will fees and costs be handled in such an event?

Figure 9-1. Other Parties with Whom Builders and Remodelers May Contract

- Accountant
- Architect
- Engineer
- Interior designer
- Experts, consultants, and specialists
  - Demolition experts
  - Environmental experts
  - Hazardous waste removal specialists (indoor and outdoor)
  - Risk management specialists

Figure 9-2. Ways Attorneys Can Assist Builders or Remodelers in Preparing Contracts

- Advise the builder or remodeler about particular provisions that should or should not be inserted in the contract and explain why, so that an astute businessperson can understand what each contract provision means at least in general terms.
- Tailor the contract to reflect state and local laws.
- Assist in drafting the contract so that it clearly and unambiguously expresses the parties' rights and obligations.
- Explain the potential legal ramifications—both positive and negative—that may arise from the contract.
- In preparing documents and in other legal matters, consult with other experts on the builder's or remodeler's behalf.
- Prepare or assist in the preparation of any documents that should accompany the contract, such as a notice of the right of rescission or a notice of the contractor's lien rights (in those states where the notice must be delivered before work is performed).
- Give advice on a regular basis. Once the attorney is familiar with the contract, he or she will be able to provide quick and accurate answers should questions arise.

In-House Real Estate and Remodeling Salespersons

Determining whether a salesperson is an employee or an independent contractor is a significant issue for builders and remodelers. At the federal level an employer is required to withhold income and Social Security taxes from employees and to pay an employer's share of Social Security. The employer is also liable for unemployment insurance tax. At the state level an employer is responsible for workers' compensation insurance for an employee.

Discussing the various tests and factors used to determine whether a particular salesperson is an employee or an independent contractor is beyond the scope of this book. Independent contractor status is determined on a case-by-case basis and depends on the true relationship of the parties. Usually how the employer acts toward the salesperson is more important to the Internal Revenue Service than what the contract states. If the working relationship can be structured to meet the independent contractor test, however, the contract should reflect that relationship.

> The salesperson provides services to the business community as an independent contractor and is engaged by [builder or remodeler] solely for the services outlined in the scope of work. Nothing in this agreement shall be interpreted or construed as creating an employer-employee relationship between subcontractor and [builder or remodeler].
>
> The salesperson is responsible for all federal and state income taxes resulting from any contract payments paid by [builder or remodeler]. The salesperson will receive an IRS form 1099 for all earnings. [Builder or Remodeler] will not have any responsibility for withholding or paying income taxes or Social Security payments on behalf of any individuals employed for work under this agreement by the salesperson.

Real Estate Brokers

In today's competitive market more new home builders are using real estate brokers to sell their homes. This relationship offers advantages to both parties: It provides the builder with access to a larger market, and it allows the broker to increase his or her inventory. A builder's use of a broker, as distinct from an in-house sales staff, would involve (a) a real estate salesperson employed by a real estate firm working on the site or (b) a builder-broker cooperative program. A builder-broker cooperative program may involve one real estate firm or many of them. In either case, however, the real estate firm does not work exclusively for the builders, and none of its employees work on the site.

A key issue under the single-firm arrangement is broker loyalty, and it should be addressed in the parties' contract. The contract should provide that any prospect inquiring by mail, telephone, or in person belongs to the builder and may not be solicited by the broker or referred by the broker to another broker. For example, if a couple walks in off the street, the broker may not offer to show the couple other properties on his or her day off, nor may the broker refer the couple to another broker.

A builder who enters into a cooperative program with outside brokers should resolve the following matters at the outset and include their decisions in their contracts with real estate brokers:

- What is expected of the broker?
- What is the commission rate? Will it be based on the home's gross sales price or on the base sales price (the price without any options)?
- Under what terms will a commission be paid? For example, if the broker's commission depends in part on him or her accompanying the prospect to and registering him or her at the builder's development upon the prospect's first visit to the site, this condition should be clearly spelled out in the contract.
- Similarly, will the builder honor the broker's registration of the prospect for a specified period of time (say 30 to 60 days)? Because the broker will be entitled to a commission during that time if the builder sells the property, a builder often will list names of his or her current active prospects in the contract with the broker and exclude sales to those prospects from commissions.
- If a dispute arises, how will it be settled? The agreement should spell out the dispute resolution procedures to be followed.
- What sales and marketing services will the real estate broker provide and which ones will the builder perform?

Some builders and brokers subscribe to builder and broker codes or guidelines that cover important matters such as those discussed in the preceding list. For example, the Southern Arizona Home Builders Association and the Tucson Association of Realtors® have adopted the Builder-Broker Code of Mutual Understanding. The builder and broker agree to subscribe to the code and to operate in accordance with its provisions.

These two groups can be contacted at the addresses listed below:

Southern Arizona Home Builders Association
2840 N. Country Club Road
Tucson, Arizona 85716
(602) 795-5114

Tucson Association of Realtors®
1622 N. Swan Road
Tucson, Arizona 85712
(602) 327-4218

Suppliers

A supplier provides builders, remodelers, and subcontractors with the materials, supplies, or equipment they use in their work. Suppliers sell building materials, and builders and subcontractors use the materials to erect and construct the a building on the site.[67] Generally a supplier differs from a subcontractor because a subcontractor has a direct contractual relationship with the builder or remodeler and undertakes a specific part of the overall work undertaken by the builder or remodeler. Usually the subcontractor agrees to provide services as well as materials or supplies.

Builders and remodelers should have written contracts with their suppliers for the same reasons that they have written contracts with their clients and their subcontractors: to document their agreements, to minimize misunderstandings, and to allocate risks.

If the parties fail to address an item in the contract or the contract does not adequately cover the item, the Uniform Commercial Code (UCC) will govern. Instead of the parties relying on the UCC, they should identify the key terms of their agreement and commit them to writing. In addition, as the contractor discovered in the case discussed below, a contractor who objects to a term in an offer from a supplier should provide an additional or different term—not merely delete the objectionable term from the acceptance—because silence may not be deemed to be a sufficient rejection.

The contract should identify the material necessary to complete the work, including accessories, and if appropriate it should note that a certain material or a certain installation is not included. If necessary, the contract should state that the products supplied and the work accomplished under the contract will conform to the plans and specifications. The material should be identified by brand name, size, quality, quantity, and other similar descriptive information. The contract should include the cost of the material (including any discounts) and the payment terms (including any deposits and retainage). Shipping terms, delivery dates, location, and conditions should be addressed in the contract. For example, if the parties have agreed that the supplier will uncrate the material and set it in place, these terms should be in the contract. The contract should provide for insurance, the return of defective and surplus materials, the supplier's warranties, and a method for resolving disputes.

Illustrative Case

Jorgensen made a written offer to sell certain goods to Mark Construction by issuing its signed quotation form, which conspicuously included a limitation of warranty liability clause. Mark accepted this offer by issuing a signed purchase order based on and including reference to Jorgensen's quotation. Mark's purchase order described the goods to be furnished, the purchase price, the terms of payment, and the time and place of delivery, as contained in Jorgensen's quotation, but the purchase order made no reference to the limitation of warranty liability provision.

When Jorgensen sued Mark for breach of contract, Mark counterclaimed against Jorgensen for breach of warranty, and Jorgensen claimed that its liability was limited under the contract. Finding for Jorgensen, the court concluded that submission of the purchase order constituted acceptance of all of the terms of the offer made in the quotation, and that Mark's silence was not an effective rejection or a counteroffer.[68]

Ten • Construction and Sales Checklist

The sample construction and sales checklists (Figures 10-1 and 10-2) provide a format for working with the home buyer or homeowner throughout the construction and sale of a home or during a remodeling project. The checklists are intended to remind builders and remodelers of the tasks that ordinarily should be performed to satisfy legal and professional requirements. The checklists should be started at the beginning of the builders' or remodelers' relationships with their buyers or homeowners and used as permanent records of the parties' actions and decisions.

The checklists merely suggest an approach to recording approvals by the buyer or the homeowner and the activities accomplished. Before using the checklists, builders and remodelers should tailor them to fit the needs of their individual businesses and construction projects.

Most of the items in Figure 10-1 will apply only when builders are (a) constructing homes on lots already owned by the home buyers, or (b) they are custom building homes. Other items in Figure 10-1 apply solely to the more common situation in which builders own the lots on which they are building. Still others apply to both situations. Some actions or methods necessary for a particular builder or product may not be included in the checklist.

A number of the items found in Figure 10-1 will also be found in the checklist for remodelers in Figure 10-2. Other items apply solely to a remodeling project. The left side of each checklist provides space to record the date and (if applicable) for the parties to initial the items. The right side of each checklist contains space for comments by the builder or remodeler. Each checklist is cross-referenced to the other parts of *Contracts and Liability for Builders and Remodelers*.

The checklist is an inexpensive way of avoiding litigation. If builders and remodelers take time before the project begins to compile a list of tasks that should be performed to satisfy legal and professional requirements and they review their lists during the course of a project, they are less likely to miss a deadline; fail to deliver a notice; forget to advise or discuss warranties, disclaimers, or other important information with the owner. For example, under the 1968 Federal Truth in Lending Act, a consumer who enters into a remodeling contract may have 3 days to rescind (terminate) the contract. The remodeler must deliver two copies of the notice of the right to rescind to each consumer entitled to rescind. The notice of rescission must be in a separate document.

If a remodeler ignores this law or does not comply fully, the contract can be canceled while the work is in progress, and the remodeler might have to refund money paid for work completed. By making the right of rescission a standard item on the checklist, the remodeler will reduce the firm's exposure to the great financial risk associated with failing to comply with this law (see Mandatory Clauses in Chapter 3).

One of the major advantages of keeping such a checklist is to assist the builder or the remodeler in the event of litigation. Cases often reach the evidence-gathering stage long after a project is completed, and the parties have difficulty recalling what happened or verifying their memories of it. A clear record kept in the form of a checklist, log, or diary will assist builders and remodelers in reconstructing events and proving, not only what they did, but also that they kept the other party informed. The existence of such a record may help to deter groundless litigation.

Figure 10-1. Sample Construction and Sales Checklist for Builders

[The builder should make up a shorter list of only the items the buyer needs to initial, and the buyer and the builder should sign that list because the buyer should see only what pertains to him or her. The buyer does not need to see the items reminding the builder of the many tasks to be done.]

Dates	Builder's Initials	Owners' or Buyer's Initials	Item	Comments
______	______		Verify that necessary state and/or local contractor's license and any other necessary legal documents are up to date.	______
______	______		Determine buyer's financial capabilities and method of payment.	______
______	______		Hold prebid conference.	______
______	______		Submit bid.	______
______	______	______	Submit initial construction documents to buyer (see Chapter 2).	______
______	______	______	Obtain certificate of title or other evidence of buyer's ownership of lot or land if appropriate.	______
______	______		Obtain performance bond (if required by buyer).	______
______	______		Obtain labor and material payment bond (if required by buyer).	______
______	______	______	Deliver bonds to buyer (if required by buyer).	______
______	______	______	Deliver certificates of insurance to buyer.	______
______	______		Make all necessary payments.	______
______	______	______	Notify buyer of list of subcontractors (see Chapter 8).	______
______	______	______	Provide buyer with estimated progress schedule.	______
______	______		Obtain copies of all property insurance policies from owner of lot or land.	______

Figure 10-1. Sample Construction and Sales Checklist for Builders (continued)

Dates	Builder's Initials	Owners' or Buyer's Initials	Item	Comments
______	______		Determine whether hazard insurance policies are necessary.	______
______	______	______	Discuss warranties on personal property with buyer. Make specific warranties available for buyer's examination (see Chapter 6).	______
______	______		Prepare construction or sales contract (see Chapter 2).	______
______	______		Review contract with attorney before executing it.	______
______	______		Receive certificate of insurance or other evidence of insurance from home buyer (including evidence of increase).	______
______	______		Review contract and all contract documents with the home buyer line by line at the contract-signing meeting (see Chapter 6).	______
______	______	______	Present builder's limited warranty and list of nonwarrantable items to buyer.	______
______	______	______	Execute contract with buyer (see Chapter 2).	______
______	______	______	Give the home buyer a list of subcontractors who will be working on the project (see Chapter 8).	______
______	______	______	Conduct preconstruction conference with home buyer; review contract, plans, specifications to ensure that home buyer has no confusion about work to be done.	______
______	______		Obtain utility (water, gas, electric) permits in lot owner's or buyer's name.	______
______	______		Submit regular (weekly, monthly) statements to buyer for payment per contract.	______

Figure 10-1. Sample Construction and Sales Checklist for Builders (continued)

Dates	Builder's Initials	Owners' or Buyer's Initials	Item	Comments
________	________	________	Consult with buyer regarding changes in plans or specifications.	________
________	________	________	Obtain home buyer's signature on written change orders with costs specified and change in completion date if any.	________
________	________		Keep dated written records of discussions with home buyer (or architect) regarding progress of work.	________
________	________		Keep dated records of any delays. (These records will prove useful if the contract contains a liquidated damages clause; see Chapter 2).	________
________	________		If a dispute arises because an incident occurs on the site, record what happened (for example, take photographs, make video recordings, have witnesses write down what they saw or heard).	________
________	________		For significant issues provide the home buyer or architect with a memo concerning the matters discussed orally.	________
________	________		Respond promptly in writing to correspondence from home buyer or architect, particularly if the correspondence accuses the builder of failing to perform in accordance with the contract. Silence may be deemed to be acquiescence.	________
________	________	________	Conduct home buyer orientation; walk through house with home buyer and fill out the Home Maintenance Checklist (see Figure 7-1).	________

Figure 10-1. Sample Construction and Sales Checklist for Builders (continued)

Dates	Builder's Initials	Owners' or Buyer's Initials	Item	Comments
________	________	________	Walk through house with home buyer and fill out punchlist. Assist home buyer in identifying all items that need improvement and establish timetable for completion.	________
________	________	________	Complete punchlist items (see Figure 7-3).	________
________	________		Have home buyer initial completed punchlist items previously listed as needing improvement (see Figure 7-3).	________
________	________		After final payment, setup appointment for any necessary warranty work and do work promptly.[1]	________
________	________		If payment is withheld file a lien; check local lien law.	________
________	________		Execute certificate of completion with home buyer (see Chapter 3).	________
________	________		Send a short letter to home buyer 30, 45, or 60 days after move-in to acknowledge the beginning of the warranty period and to solicit a list of warranty items that need attention. Alternatively arrange a follow-up meeting with the home buyer 4 to 6 weeks after move-in.	________
________	________		Shortly before the end of the warranty period advise the buyer in writing that the warranty is nearing expiration and invite the owner to submit a final list of items that require warranty attention.[2]	________

1. See also Carol Smith, *Warranty Service for Builders and Remodelers* (Washington, D.C.: Home Builder Press, National Association of Home Builders, 1991) pp. 92–96.
2. Smith, pp. 92–96.

Figure 10-2. Sample Construction and Sales Checklist for Remodelers

[The remodeler should make up a shorter list of only the items the owner needs to initial, and the remodeler and the owner should sign that list because the buyer should see only what pertains to him or her. The owner does not need to see the items reminding the remodeler of the many tasks to be done.]

Dates	Remodeler's Initials	Owners' or Homeowner's Initials	Item	Comments
______	______		Verify that necessary state and/or local contractor's license and any other necessary legal documents are up to date.	______
______	______		Determine whether the homeowner actually owns the property. Check local land records. Request certificate of title from owner's title insurance company.	______
______	______		Determine the homeowner's financial capabilities and methods of payment.	______
______	______	______	Educate the homeowner on the nature of remodeling and the company's method of doing business.	______
______	______		Hold prebid conference.	______
______	______		Submit proposal.	______
______	______	______	Submit initial construction documents to homeowner.	______
______	______	______	Deliver certificate of insurance or other evidence of insurance to homeowner.	______
______	______	______	Provide homeowner with estimated progress schedule.	______
______	______	______	Discuss manufacturer's warranties on personal property with homeowner. Make specific warranties available for homeowner's examination (see Chapter 6).	______
______	______		Consult with the homeowner regarding plans and specifications,	______

Figure 10-2. Sample Construction and Sales Checklist for Remodelers (continued)

Dates	Remodeler's Initials	Owners' or Homeowner's Initials	Item	Comments
________	________	________	Present remodeler's limited warranty and list of nonwarrantable items to homeowner.	________
________	________		Prepare a construction contract (see Chapter 3).	________
________	________		Review contract with attorney before executing it.	________
________	________		Receive certificate of insurance or other evidence of insurance from homeowner (including evidence of increase).	________
________	________		Review contract and all contract documents with the homeowner line by line at the contract-signing meeting.	________
________	________		Execute contract with homeowner (see Chapter 3).	________
________	________		Give the homeowner any required notices, such as the notice of rescission or the notice of the remodeler's lien rights (see Chapter 3).	________
________	________	________	Give the homeowner a list of subcontractors who will be working on the project (see Chapter 8).	________
________	________		Conduct preconstruction conference with homeowner; review contract, plans, specifications to ensure that homeowner has no confusion regarding work to be done.	________
________	________		Obtain necessary permits.	________
________	________		Submit regular statements (tied to contract terms) to homeowner for payment.	________

Figure 10-2. Sample Construction and Sales Checklist for Remodelers (continued)

Dates	Remodeler's Initials	Owners' or Homeowner's Initials	Item	Comments
________	________		Consult with homeowner regarding changes in plans or specifications.	________
________	________		Obtain labor and material payment bond (if required by owner).	________
________	________		Obtain homeowner's signature on written change orders with costs specified and new completion dates if necessary.	________
________	________		Keep dated written records of discussions with homeowner (or architect) regarding progress of work.	________
________	________		Keep dated records of any delays. (These records will prove useful if the contract contains a liquidated damages clause (see Chapter 3).	________
________	________		If a dispute arises because an incident occurs on the site, record what happened (for example, take photographs, make video recordings, have witnesses describe what they saw or heard in writing).	________
________	________		For significant issues, provide the homeowner or architect with a hand-written memo concerning the matters discussed orally).	________
________	________		Respond promptly in writing to correspondence from homeowner or architect, particularly if the correspondence accuses the remodeler of failing to perform in accordance with the contract. Silence may be deemed to be acquiescence.	________
________	________	________	Conduct home buyer orientation; walk through house with homeowner and fill out the Home Maintenance Checklist (see Figure 7-1).	________

Figure 10-2. Sample Construction and Sales Checklist for Remodelers (continued)

Dates	Remodeler's Initials	Owners' or Homeowner's Initials	Item	Comments
______	______	______	Walk through house with homeowner and fill out punchlist. Assist homeowner in identifying all items that need improvement and establish timetable for completion.	______
______	______	______	Complete punchlist items (see Figure 7-3).	______
______	______		Have homeowner initial completed punchlist items previously listed as needing improvement (see Figure 7-3).	______
______	______		After final payment, set up appointment for any necessary warranty work and do work promptly.[1]	______
______	______		If payment is withheld file a lien; check local lien law.	______
______	______		Execute certificate of completion with homeowner (see Chapter 3).	______
______	______		Send a short letter to homeowner 30, 45, or 60 days after move-in to acknowledge the beginning of the warranty period and to solicit a list of warranty items that need attention. Alternatively arrange a follow-up meeting with the owner 4 to 6 weeks after move-in.	______
______	______		Shortly before the end of the warranty period advise the owner in writing that the warranty is nearing expiration and invite the owner to submit a final list of items that require warranty attention.[2]	______

1. See also Carol Smith, *Warranty Service for Builders and Remodelers* (Washington, D.C.: Home Builder Press, National Association of Home Builders, 1991) pp. 92–96.
2. Smith, pp. 92–96.

Appendixes. Sample Agreements

Warning—The sample contracts that follow are merely provided for educational purposes to illustrate the principles discussed in this book and are not to be used as forms. The contracts are designed to cover the major areas of consideration for most new residential construction contracts and remodeling contracts. However, the suggested contract provisions do not and cannot apply to every situation, nor do they comply with any particular state law. Some of the provisions will not apply to a particular situation while in other cases additional terms may be appropriate. Builders and remodelers should work with their attorneys to prepare documents that meet their particular needs.

Appendix A. Sample Contract for Construction of Single-Family Home on Owner's Lot

This agreement is made this ____________ day of ___________, 19___, by and between XYZ Builders, Inc. (hereinafter referred to as the builder), and Jane Kay Doe and John Lee Doe (hereinafter referred to as the owners) for the construction of a single-family residence for the owners on the property located in ________________ County, State of ________________ and legally described as lot _______ block ________ subdivision ____________________ and also known and numbered as ____.

The builder and the owners agree as set forth below:

1. Contract Documents—The terms of this contract include the conditions of this contract and by reference the provisions in the other documents specifically listed below. (Copies of these documents are attached to this contract as Exhibits A, B, C, D, and E.) The terms of this agreement shall prevail over any conflicting provisions in the documents incorporated by reference. If a conflict exists between the plans and the specifications, the specifications shall govern.

Contract documents:

	Title	Date	No. of Pages	Signed
A.	______________________	________	________	________
B.	______________________	________	________	________
C.	______________________	________	________	________
D.	______________________	________	________	________

2. The Work—Unless otherwise specifically noted, the builder shall provide and pay for all labor, materials, equipment, tools, construction equipment and machinery, transportation, and other facilities and services necessary for the proper execution and completion of the residence. The work shall be done substantially in conformance with the plans and specifications that have been initialed or signed by each party. These plans and specifications are attached to this contract and incorporated into it as Exhibit A. Because the plans and specifications were prepared by the owners' architect, the parties agree that the builder shall not be liable for errors or omissions attributable to the plans and specifications when the builder has complied with them.

The builder shall not be responsible for the following work: hauling excavation and existing debris from the property; off-site utility trenches; installation or construction of walks, pavements, or curbing; installation of cable television wiring; interior painting of garage; and extra work caused by the presence of concealed conditions as set forth in paragraph 12.

3. Financing—This contract is contingent upon the owners obtaining a construction loan in the amount of ___________________________ dollars ($______________). All fees and expenses of obtaining a loan including all commissions, title charges, and credit reports shall be borne by the owners. The builder is not required

_____/_____
Owner's Initials

to begin construction until the owners provide the builder with written notice from the lender that the owners have closed on said loan. If the owners cannot obtain financing within thirty (30) days from the date the builder approves this contract, either party has ten (10) days thereafter to elect to terminate this contract by giving written notice to the other party. The builder shall refund to the owners all money paid, less costs and obligations incurred by the builder at the owners' request.

4. Contract Price—The owners agree to pay the total contract price for all labor and materials furnished and work performed by the builder of ______________________ dollars ($____________), including (name of state) state sales tax, subject to additions and deletions by change order as provided in paragraph 11. The contract price includes the allowances listed in the Allowance Schedule attached to this contract and incorporated into it as Exhibit B.

The allowance includes both materials and installation unless expressly noted otherwise. The parties agree that the allowances are not to be construed as bids by the builder and that the allowances may vary from the actual cost based on the owners' elections. If the cost of the owner-selected materials or their installation exceeds the material or installation allowance, the amount of that excess will be added to the next progress payment or the final payment. If the amount is less than the allowance amount, that amount will be subtracted from the final amount of the contract.

5. Payments—The contract price will be paid as follows:

A. $______________ as a deposit upon signing the contract, receipt of which is hereby acknowledged.

B. Based on applications for payment submitted by the builder, the owners shall make progress payments toward the contract price in accordance with the Construction Draw Schedule of the owners' construction lender as work is completed and certified by the owners' construction lender. The owners will pay the cost of each inspection for each draw request. (The Construction Draw Schedule is attached to this contract and incorporated in it as Exhibit C.)

C. The owners agree to make the progress payments within five (5) days of certification by the owners' construction lender. Payments due and unpaid shall bear interest (at the maximum legal rate) payable to the builder from the date the payment is due. If the owners fail to pay the builder within seven (7) days of the date the payment is due through no fault of the builder, upon three (3) additional days' written notice to the owners, the builder may stop the work. The builder may keep the job idle until such time as payments that are due to the builder are paid.*

6. Acceptance, Final Payment, and Occupancy—Upon receipt of written notice that the work is ready for final inspection and acceptance, the owners will promptly inspect the work. When the owners' lender finds the work acceptable under the contract ("substantial completion"), the owner will promptly pay (or cause to be paid) the balance due under the contract less an amount equal to the cost to complete any missing or unfinished remaining punchlist items (the punchlist escrow).

The builder agrees to provide the owners with (a) an affidavit stating that all materials and services for which a lien could be filed have been paid or will be paid from the proceeds and (b) such other affidavit as may be reasonably required by the owners' title insurance company. The punchlist escrow will be paid to the builder immediately upon completion of each of the punchlist items. Occupancy will be granted to the owner when construction is substantially completed, the certificate of occupancy is issued, and the builder receives payment

*Some builders establish a special checking account into which the lender deposits the draws, so the owners never see or touch the money involved.

_____/_____
Owner's Initials

of the final draw (including payment for all change orders and overages of allowances), less any money held for incomplete items.

7. Commencement and Completion—The work will begin promptly after the owners have obtained the financing referred to above, any other contingencies are cleared, and the owner provides the builder with a title report containing a correct statement of (a) the recorded legal title to the property on which the residence is to be built and (b) the owners' interest therein at the time of the signing of this contract. The work will be substantially completed within one hundred and eighty days (180) from the date the building permit is obtained.

Any time lost by reason of change in plans or specifications requested by the owners, other acts of the owners, strikes, weather conditions not reasonably anticipated, or any other conditions that are not within the builder's control shall be added to the specified time of completion. And the builder shall not be liable for such delay. For any delays not the responsibility of the builder, the contract price shall increase by the difference, if any, in the builder's costs occasioned by such delay. A claim for an increase in the time for performance of the contract shall be made within ____________________ (_____) days after the occurrence of the event giving rise to the claim or within ____________________ (_____) days after the builder first recognizes the condition giving rise to the claim, whichever is later.

8. Selections—Upon signing this contract, the builder will give the Selection Guide (attached to this contract and incorporated in it as Exhibit D) to the owners to help them select allowance items, materials, and colors required during the construction process. Exterior materials (brick, shingles, siding, and so on) selections must be made within twenty one (21) days. The remaining selections must be made within thirty (30) days of the signing of this contract to avoid delays.

9. Permits, Fees, and Tests—The builder shall secure and pay for building permits, licenses, and other similar approvals necessary for the proper execution and completion of the work. If necessary, the owners agree to assist the builder in obtaining any such permits and licenses by completing all necessary applications and forms. However, if a covenant or an architectural review committee requires the approval of plans and specifications, the owners shall be responsible for obtaining these approvals and paying for any fees connected with them. If no soil report is currently available, the owners shall provide one at their expense.

10. Taxes—The owners shall pay all real property taxes and taxes imposed upon the improvements on the residence when they are due. The builder shall pay all necessary sales, use, and similar taxes on materials used in construction that are legally enacted at the time this contract is signed.

11. Change Orders—Without invalidating this contract, the owners may order changes in the work within the general scope of the contract. However no changes are to be made except upon a prior written order (signed by both parties) consisting of the change, any additional cost, and the additional number of days to be added to the completion date.

If the change reduces the cost, the owners will receive a credit, but the builder's supervision and overhead expenses and profit will not be reduced. Any additional cost shall be paid for prior to installation, and the construction loan account may not be used to pay for changes. The owners agree to make requests concerning any changes, additions, or alterations in the work to the builder, and the owners agree not to issue any instructions to, or otherwise negotiate for additional work with, the builder's subcontractors or employees. Either owner may sign the change order on his or her own behalf and on behalf of the other, and the signature shall be binding on both owners.

12. Insurance—The builder shall purchase and maintain at the builder's own expense, all necessary workers' compensation and employer's liability insurance, commercial general liability insurance, and compre-

_______/_______
Owner's Initials

hensive automobile liability insurance to protect the builder from claims for damages because of bodily injury, including death, and for damages to property that may arise both out of and during operations under this contract.

The owners shall purchase and maintain their own liability insurance, including fire and casualty insurance upon the residence, to the full insurable value and shall name the builder as an additional insured. Each party shall issue a certificate of insurance to the other prior to construction.

13. Owners' Obligations—The owners shall (a) furnish all surveys describing the physical characteristics, and utility locations for the residence and (b) secure and pay for easements necessary for the completion of the work. The owners shall furnish information and services under the their control to the builder promptly to avoid delay.

The owners warrant that the property upon which the residence is to be built conforms to all zoning, planning, environmental, and other building requirements. The owners warrant that all utilities necessary for the completion of construction are to the property line.

14. Concealed Conditions—The builder has visited the site and has familiarized itself with the local conditions under which the work is to be performed. However the builder is not responsible for subsurface or latent physical conditions at the site or in an existing structure that differ materially from those (a) indicated or referred to in the contract documents or (b) ordinarily encountered and generally recognized as inherent in the work of the character provided for in this contract.

After receiving notice of the condition, the owners shall investigate the condition within five (5) business days. If the parties agree that the condition will increase (a) the builders cost of performance or any part of the work under this contract or (b) the time required for that work, the parties may sign a change order agreement incorporating the necessary revisions, or the owners may terminate the contract.

If the owners terminate the contract the builder will be entitled to recover from the owner payment for all work performed, including normal overhead, and a reasonable profit.

15. Disputes—Should any dispute arise relative to the performance of this contract that the parties cannot satisfactorily resolve, if the parties so agree, the dispute shall be referred to a single arbitrator acceptable to them. If the parties do not so agree, then the parties agree that the dispute shall be resolved by binding arbitration conducted by the American Arbitration Association. The party demanding arbitration shall give written notice to the opposite party and the American Arbitration Association promptly after the matter in dispute arises. In no event, however, shall a written notice of demand for arbitration be given after the date on which a legal action concerning the matter in dispute would be barred by the appropriate statute of limitations.

16. Termination—If the builder fails to supply proper materials and skilled workers; make payments for materials, labor, and subcontractors in accordance with their respective agreements; disregards ordinances, regulations, or orders of a public authority; or fails to materially comply with the provisions of the contract, the owners may give the builder written notice to terminate. After seven (7) days if the builder has failed to remedy the breach of contract, the owner can give a second notice to terminate. If the builder still fails to cure the breach within three (3) days after the second notice, the owners may terminate the contract.

17. Limited Warranty—Upon receipt of the final payment in full, the builder shall deliver to the owners the XYZ, Inc., Limited Warranty Agreement. This agreement provides warranties subject to stated terms and conditions against certain kinds of defects in workmanship and materials for the time stated therein. A copy of the Limited Warranty is attached to this contract and incorporated in it as Exhibit E.

_______/_______
Owner's Initials

This Limited Warranty is the only express warranty provided by the builder. Implied warranties, including (but not limited to) warranties of merchantability, fitness for a particular purpose, habitability, and good workmanship are limited to the warranty period set forth in the XYZ, Inc., Limited Warranty Agreement.

The owners acknowledge that the builder has made no guarantees, warranties, understandings, nor representations (nor have any been made by any representatives of the builder) that are not included in the contract documents. _____/____
Owner's initials

18. Miscellaneous—Until the owners make the final payment and take possession, the owners agree that the builder shall have the right to place signs on or about the property and to show the residence to other prospective clients and customers.

19. Governing Law and Assignment—This contract will be construed, interpreted, and applied according to the law of the state where the property is located. This contract shall not be assigned without the written consent of all parties.

20. Effective Date and Signature—This contract shall become effective on the day it is signed by both parties.

We the undersigned, have read, understand, and agree to each of the provisions of this contract and hereby acknowledge receipt of a copy of this contract.

XYZ Builders, Inc.

By ______________________________ ______________________________
Title Date

By ______________________________ ______________________________
Jane Kay Doe, Owner Date

By ______________________________ ______________________________
John Lee Doe, Owner Date

Contract Documents
Final Contract
Exhibit A. Plans and Specifications
Exhibit B. Allowance Schedule
Exhibit C. Construction Draw Schedule
Exhibit D. Selection Guide
Exhibit E. XYZ, Inc., Limited Warranty Agreement

Appendix B. Sample Remodeling Contract

This contract is made between XYZ Remodelers, Inc., whose principal place of business is at

____
(street)

______________________ ________ ____________
(city) (state) (zip)

Jane Kay Doe and John Lee Doe, whose address is

____
(street)

______________________ ________ ____________
(city) (state) (zip)

1. Contract Documents—The contract documents include the terms of this remodeling contract and, by reference, the provisions of the other documents listed below. (Copies of these documents are attached to this contract.) The contract, plans, and specifications are intended to supplement each other. If these documents conflict, however, the specifications shall control the plans, and the contract shall control both.

Contract documents:

	Title	Date	No. of Pages	Signed
A.				
B.				
C.				
D.				

2. Description of Work—The remodeler will furnish all labor, materials, construction equipment and machinery necessary to complete certain alterations and improvements in accordance with the plans and specifications upon the following described property:______________________________.

This contract does not include the following tasks:

A. Changing electrical service or any other electrical equipment that is in violation of applicable electrical or building codes.
B. Painting of the existing house interior or exterior.

_____/_____
Owner's Initials

C. The identification, detection, abatement, encapsulation, handling, or removal of any hazardous material as defined by the Environmental Protection Agency (EPA) that is present on the site prior to the time of commencement of the work described in this contract. The remodeler reserves the right to stop work until such materials are properly removed.

3. Contract Price—In consideration for the work to be performed by the remodeler, the owners shall pay the remodeler the sum of ______________________________ Dollars ($________________) (subject to additions and deductions by written change order). This contract price is based on the cost of the work as defined by the plans and specifications. The cost of the work not specifically defined by the plans and specifications has been added into the contract price by budgeting an allowance for each item. A list of these allowances is attached to this contract and is incorporated into it by reference. If the owners' selections and/or actual construction costs result in costs exceeding the allowances, the owners agree to pay the remodeler such additional costs. If the Owners' selections and/or actual costs result in costs less than the allowances, the remodeler agrees to credit the owners for the difference. The contract price shall be payable as follows:

$ __________ deposit at contract signing
$ __________ at start of ______________________________
$ __________ at start of ______________________________
$ __________ at start of ______________________________
$ __________ when ready for ____________________________
$ __________ when ready for ____________________________
$ __________ when ready for ____________________________
$ __________ substantial completion (Substantial completion occurs when the work is suitable for its intended use, or the Building Department issues an occupancy permit, whichever occurs first).

4. Final Payment, Inspections, and Liens—Upon notification by the remodeler of substantial completion of the work, the owners and the remodeler shall inspect the work performed and at that time the owners shall prepare a punchlist that identifies any deficiencies in workmanship or materials. The owners may retain the value of the punchlist work from the final payment until the punchlist items are complete. Upon completion of the punchlist items (not to exceed ______________ (_____) days from the date of the punchlist), the owners shall pay the remodeler the balance of the contract price. At final payment the remodeler shall deliver to the owners a release of lien in the amount then due. The remodeler shall furnish lien releases with each request for and receipt of payment.

5. Application for Payment—The remodeler shall prepare itemized statements for the owners when payments are due under the provisions of paragraph 3 above. All payments are due within 10 days of the date the remodeler submits each payment request. Payments due and unpaid shall bear interest at the rate of two percent (2%) per month. The parties agree that this late charge represents a fair and reasonable estimate of the costs the remodeler will incur by reason of the late payment by the owners. If any payment is not paid when due, if the remodeler has provided a written demand to the owners for payment, and after seven (7) days the owners have failed to make the payment, the remodeler may suspend the work until such time as all payments due have been paid.

_______/_______
Owner's Initials

6. Time of Performance and Financial Arrangements—The remodeler shall commence work as soon as practical after signing this contract, but not until the owners have made financial arrangements that are satisfactory to the remodeler. The remodeler may at its option revoke this contract if the owners have not acquired satisfactory financial arrangements within thirty (30) days from the date of this contract. The owners agree to provide the remodeler with a signed copy of a commitment for the financing of the work prior to commencement of construction. The remodeler shall substantially complete all work under this contract within ________________ (_____) working days after commencement, subject to the following delays: failure to obtain all necessary building permits and other approvals within a reasonable length of time, acts of God, weather, strikes, owners' change orders, acts of neglect or omission by owner, changes caused by inspectors of governmental agencies, or other causes beyond the remodeler's control.

If the remodeler is delayed at any time in the progress of the work as described above, the completion schedule for the work or the affected parts of the work shall be extended by the same amount of time caused by the delay.

7. Owners' Responsibilities—The owners shall secure and pay for any easements, variances, zone changes, necessary modifications of restrictive covenants, or other actions. The owners will indicate the property lines to the remodeler and will provide boundary stakes by a licensed land surveyor if the owners are in doubt about the property boundaries. The owners assume all responsibility for the accuracy of the boundary markers. The owners agree to make drinking water and toilet facilities available to the remodeler, the remodeler's employees, and subcontractors or to pay the remodeler for the cost of rented units.

The owners shall pay for assessments, hook-up charges, connection or tapping fees required by public bodies or utilities and for electricity, water, gas, sewers, and other utilities required for the work to be performed. The owners shall permit the remodeler and the remodeler's workers access to a telephone for local business calls related to the job. No long distance calls will be made from the owners' home.

8. Permits, Licenses and Other Approvals—The remodeler shall obtain and pay for all local building and construction permits, licenses, governmental charges and inspection fees, and all other approvals necessary for the work, occupancy of permanent structures, or changes in existing structures that are applicable when the contract is signed, except as provided otherwise in this contract.

9. Change Orders—During the course of the project, the owners, without invalidating the contract, may order changes in the work consisting of additions, deletions, or modifications. (The contract sum and the contract time will be adjusted accordingly.) All such changes in the work shall be authorized by a written change order signed by the parties. The owners agree that changes resulting in the furnishing of additional labor or materials shall be paid for prior to commencement of the extra work. The owners agree that, to expedite the work, either of them may sign the change order, and that the signature of one is binding on the other.

10. Uncovering and Correcting Work—If the work is covered in contradiction to the contract or the applicable laws, the remodeler shall uncover the work for observation or inspection and replace it without charge unless the owners caused the problem that needs to be corrected.

Upon written notice by the owners, the remodeler shall promptly redo and recover at its own expense, the work that does not meet contract specifications.

11. Access—The owner shall grant free access to work areas for workers and vehicles and shall allow areas for storage of materials and debris. Driveways shall be kept clear and available for movement of vehicles during the scheduled working hours, which will be ______ a.m. to ______ p.m., Monday through Saturday. The

_______/_______
Owner's Initials

remodeler and the remodeler's workers shall make reasonable efforts to protect driveways, lawns, shrubs, or other vegetation. However they shall not be responsible for damage to any of the items listed above unless the damage results from their gross negligence. If off-street parking is not available, the owners agree that the remodeler will secure parking permits for on-street parking as needed, and the cost of the permits will be included in the final contract price.

12. Matching Materials—The remodeler calls attention of the owners to the limitations of plaster, stucco, concrete, and so on, and while the remodeler will make every effort to blend existing textures, colors, and planes, exact duplication is not guaranteed. Roofing materials and/or the color of the roofing material may not match existing roofing because the manufacturer discontinued the color or product, shading difference of product, age, or changes in the product or the product's manufacturing process.

13. Differing Site Conditions—The remodeler stipulates that it has visited the home site, and it has familiarized itself with the local conditions under which the work is to be performed. However the owners acknowledge that this contract is based solely on observations that the remodeler was able to make with the structure in its current condition at the time the work was bid, and that the remodeler shall not be responsible for differing site conditions that were not visible when this contract was made unless stated otherwise in the specifications.

Differing site conditions may include, but are not limited to, damage caused by termites or dry rot, rock not removable by ordinary hand tools, hidden pipes, and code violations that must be repaired, corrected, replaced, or overcome. Before disturbing any such differing site condition, the remodeler shall, if possible and reasonably practical, notify the owners of such condition, and the parties shall execute such reasonable change orders as may be appropriate in the circumstances or terminate the contract.

14. Insurance and Risk of Loss—During the term of this contract, the owners shall maintain insurance covering the full replacement cost of the improvement under contract, theft of materials on site, and any damage to persons or property arising out of the work. The remodeler shall purchase and maintain such insurance as is necessary to protect the owners from claims under workers compensation and from liability to others for damages because of bodily injury, including death, and from liability for damages to property. If the work, in whatever stage of completion, is destroyed or damaged by any accident, casualty, disaster, or calamity, including, but not limited to, fire, storm, flood, or earthquake, the owners shall be responsible for such loss, and if the estimated cost to rebuild or restore the work already performed is less than ____________ percent (_____%) of the contract price specified in paragraph 3 above, the remodeler shall prepare a written change order for the labor, materials, and profit and overhead costs required to repair the damage.

Upon signing of the change order by the owners, the remodeler shall proceed to repair the damage. The cost of the change order shall be paid from the proceeds of insurance and funds of the owners. Subsequent to completion of the repairs, the remodeler shall continue to work under the original contract. If the estimated cost to rebuild or restore the work already performed exceeds _________ percent (_____%) of the contract price specified in paragraph 3 above, the owners shall have the option of canceling this contract, in which case the owners shall pay the remodeler for all work completed prior to the notice of termination, including the remodeler's overhead, plus a net profit of _________ percent (_____%) of the full contract price.

15. Clean-Up—The remodeler will leave the work site orderly at the end of each day. Upon completion of the work, all of the remodeler's construction debris and equipment shall be removed by the remodeler and the premises left in neat, broom-clean condition, unless the parties make some other arrangement.

_____/_____
Owner's Initials

16. Dispute Resolution—The owners and the remodeler will cooperate with one another in avoiding and informally resolving disputes between them. However all claims or disputes arising out of, or relating to this contract that are not informally resolved shall be resolved by mediation or arbitration provided by ________________________, under their rules and procedures, and the decision by the arbitrator shall be final and binding on the parties. The arbitrator shall award reasonable costs and expenses, including attorney's fees, to the prevailing party.

17. Entire Agreement—This contract (including the documents incorporated by reference) constitutes the entire agreement between the parties and supersedes all prior or contemporaneous agreements and understandings between the parties whether oral or written. This contract shall not be assigned without the written consent of all parties.

This contract has no force or effect and will not be binding upon the remodeler until it is accepted and signed by the owners and countersigned by the remodeler.

We the undersigned, have read and understand and agree to each of the provisions of this contract and hereby acknowledge receipt of a copy of this contract.

XYZ Remodelers, Inc.

By ________________________________ ________________________________
Title Date

By ________________________________ ________________________________
Jane Kay Doe, Owner Date

By ________________________________ ________________________________
John Lee Doe, Owner Date

Notes

Chapter 1. Reducing Liability

1. *Quality Standards for the Professional Remodeler*, 2nd ed. (Washington, D.C.: Home Builder Press, National Association of Home Builders, 1991), 81 pp.
2. *Kaiser v. Fishman*, 187 A.D.2d 623, 590 N.Y.S2d 230 (1992).
3. Builders, remodelers, or local affiliates of NAHB interested in establishing such a system should consult *An HBA Consumer Relations Plan: A Guide to NON-HOW Complaint Handling Systems* (Washington, D.C.: National Association of Home Builders, 1978).
4. *Quagliana v. Exquisite Home Builders, Inc.*, 538 P.2d 301 (Utah 1975).
5. *Gamble v. Main*, 171 W. Va. 469, 300 S.E.2d 110 (1983).
6. *Strawn v. Canuso*, 657 A.2d 420 (N.J. 1994).
7. *ABC Builders, Inc. v. Phillips*, 632 P.2d 925 (Wyo. 1981).
8. *Kriegler v. Eichler Homes, Inc.*, 269 Cal. App.2d 224, 74 Cal. Rptr. 749 (1969).
9. *Certain-Teed Products Corp. v. Goslee Roofing and Sheet Metal, Inc.*, 339 A.2d 302 (Md. App. 1975).

Chapter 2. Contract Between Builder and Buyer or Owner

10. *Batter Building Materials Co. v. Kirschner*, 110 A.2d 464 (Conn. 1954).
11. *Surety Development Corp. v. Grevas*, 42 Ill. App.2d 268, 192 N.E.2d 145 (1963).
12. *Winn v. Aleda Construction Co., Inc.*, 315 S.E.2d 193 (Va. 1984).
13. McNeill Stokes, *Construction Law in Contractors' Language*, 2nd ed. (N.Y.: McGraw-Hill, 1990), p. 54 .
14. K. Collier, *Construction Contracts* (Englewood Cliffs, N.J.: Prentice Hall, 1987), p. 151.
15. *Naylor v. Siegler*, 613 S.W.2d 546 (Tex. Civ. App. 1981).
16. *Grubb v. Cloven*, 601 S.W.2d 244 (Tex. Civ. App. 1981).
17. Steven Stein, *Construction Law*, vol. 1. ¶5.07[1] [b][v] (New York: Matthew Bender, 1995).
18. *Old Post Office Plaza Limited Partnership v. Goodwin*, No. CVN-8910-889, 1991 WL 270281 (Conn. Sup. Ct. Nov. 18, 1991).
19. *V. L. Nicholson Co. v. Transcon Investment and Financial Ltd., Inc.*, 595 S.W.2d 474 (Tenn. 1980).
20. Maryland Real Property Code Annotated, §10-506(a) (1988).
21. See 16 Code of Federal Regulations §460.16 (1992).
22. *Hudson v. D & V Mason Contractors, Inc.*, 252 A.2d 166 (Del. Super. Ct. 1969).
23. Jerry Householder and John C. Mouton, *Estimating Home Builders* (Washington, D.C.: Home Builder Press, 1992), pp. 77–79.
24. James Acret, *Construction Litigation Handbook* (Colorado Springs: Shepards/McGraw-Hill, 1986) p. 78.

Chapter 3. Contract Between Remodeler and Owner

25. Case, Linda W, *Remodelers Business Basics* (Washington, D.C.: National Association of Home Builders, 1989), p. 19.
26. *Sites v. Moore*, 79 Ohio App.3d 694, 607 N.E.2d 1114 (1992).
27. *Walsh Services v. Feek*, 45 Wash.2d 289, 274 P.2d 117 (1954).
28. McNeil Stokes, *Construction Law in Contractor's Language*, 2nd ed. (New York: McGraw-Hill, 1990), p. 311–12.
29. K. Collier, *Construction Contracts*, Englewood Cliffs, N.J.: Prentice Hall, 1987), p. 151.
30. *Hanrahan v. Audubon Builders, Inc.*, 614 A.2d 748 (Pa. Super. 1992).
31. *Welch v. Fuhrman*, 496 So.2d 484 (La. App. 1986).
32. Steven Stein, *Construction Law*, vol. 1, ¶5.07[1][b] [v] (New York: Matthew Bender, 1995).
33. *Martin v. Phillips*, 440 A.2d 1124 (N.H. 1982).
34. *Old Post Office Plaza Limited Partnership v. Goodwin*, No. CVN-8910-889, 1991 WL 270281 (Conn. Sup. Ct. Nov. 18, 1991).
35. *Kaufman v. Gray*, 135 A.2d 455 (D.C. 1957).
36. McNeil Stokes, *Construction Law in Contractor's Language*, 2nd ed. (New York: McGraw-Hill, 1990), p. 192.
37. *Caulkins v. Petrillo*, 200 Conn. 713, 513 A.2d 43 (1986).
38. Texas Property Code Annotated, §41.007 (Watsonville, Calif.: Merk Publishers, 1992).
39. See also *A Primer for Builders...New Home Warranties and the Magnuson-Moss Warranty Act* (Washington, D.C.: National Association of Home Builders, 1978).
40. See 16 Code of Federal Regulations, Part 429 (1992) (Cooling Off Period for Door-To-Door Sales).
41. See 12 Code of Federal Regulations, Part 226 (1992), (Regulation Z, Truth in Lending); *What Builders and Remodelers Should Know About Right of Rescission Provision in the Truth in Lending Act* (Washington D.C.:

Consumer Affairs Dept., National Association of Home Builders, 1987).

42. *Einhorn v. Ceran Corporation*, 177 N.J. Super. 442, 426 A.2d 1076 (1980).
43. *Denice v. Spotswood I. Quinby, Inc.*, 248 Md. 428, 237 A.2d 4 (1968).
44. Adapted from "Contract for Repairs or Alterations," Clause 8, model contract (Louisville, Ky.: Home Builders Association of Louisville, 1993).
45. Adapted from "Terms and Conditions," Clause 13 model contract (San Antonio, Texas: Greater San Antonio Builders Association, 1981).

Chapter 4. Design-Build Contracts Used by Remodelers and Custom Builders

46. For a discussion of this topic see Linda W. Case, *Design/Build for Remodelers, Custom Builders, and Architects* (Washington, D.C.: Home Builder Press, National Association of Home Builders, 1989) 149 pp.
47. See Christopher C. Whitney, "An Evolving Perspective on Design/Build Construction: A View from the Courthouse," *The Construction Lawyer*, Vol. 15, No. 2 (April 1995), p. 97, n. 29.

5. Environmental Liability in Real Estate Transactions

48. 42 United States Codes [Typesetter insert pound sterling mark] 9601 (35) (B) (1986).
49. *Tanglewood East Homeowners v. Charles-Thomas, Inc.*, 849 F.2d 1568 (5th Cir. 1988).
50. See the EPA's main public information pamphlet on radon, *A Citizen's Guide to Radon: What It Is and What To Do About It*, 2nd ed. (Washington, D.C.: U.S. Environmental Protection Agency (EPA), U.S. Department of Health and Human Services, and U.S. Public Health Service, 1992), p. 13. Available from EPA.
51. *A Citizen's Guide to Radon*, p. 9.
52. *Radon Reduction in New Construction: An Interim Guide*, OPA-87-009 (Washington, D.C: U.S. Environmental Protection Agency, 1987).
53. 40 Code of Federal Regulations, §61.141 (National Emission Standard for Asbestos)
54. *What Remodelers Need to Know and Do About Lead* (Washington, D.C.: Technology and Codes Department and Remodelers® Council, National Association of Home Builders, 1992,), 15 pp.

6. Warranties and Disclaimers

55. *Bridges v. Ferrell*, 685 P.2d 409 (Okla. App. 1984).
56. *Quality Standards for the Professional Remodeler* (Washington, D.C.: Home Builder Press, National Association of Home Builders, 1991), 81 pp.
57. *Breckenridge v. Cambridge Homes, Inc.*, 246 Ill. App.3d 810, 616 N.E.2d 615 (1993).
58. See *A Primer for Builders . . . New Home Warranties and the Magnuson-Moss Warranty Act* (Washington, D.C.: National Association of Home Builders, 1978), available from the NAHB Consumer Affairs Department.
59. See "2. Exclude consumer products," *A Primer for Builders . . .*, p. 6.
60. See "2. Exclude consumer products," *A Primer for Builders . . .*, p. 6.
61. See "3. Conform whole warranty," *A Primer for Builders . . .*, p. 6.

Chapter 7. Inspections

62. See Carol Smith, *The Positive Walk-Through* (Washington, D.C.: National Association of Home Builders, 1990), 139 pp.

Chapter 8. Subcontract

63. Adapted from N. Schemm, "Subcontract Forms from the Subcontractor's Perspective." Reprinted with the permission of *The Practical Real Estate Lawyer*. Subscription rates $35/year; $8.75/issue. This article appeared in the September 1986 issue of *The Practical Real Estate Lawyer*.
64. *Slaught v. Bencomo Roofing Co.*, 30 Cal. Rptr. 2d 618, 25 Cal. App. 4th 744 (1994).
65. *Industrial Indemnity Co. v. Wick Construction Co.*, 680 P.2d 1100 (Alaska 1984).
66. *J. M. Process Systems, Inc., v. W. L. Thompson Electric Co.*, 218 Ill. App. 3d 350, 578 N.E.2nd 264, 267 (1991).

Chapter 9. Contracts with Other Team Members

67. *Duluth Steel Fabricators, Inc. v. Commissioner of Taxation*, 237 N.W.2d 625 (Minn. 1975).
68. *Earl M. Jorgensen Co. v. Mark Construction, Inc.*, 540 P.2d 978 (Hawaii 1975).

Selected Bibliography

Acret, James. *Construction Litigation Handbook.* Colorado Springs: Shepards and McGraw-Hill, 1986. 477 pp. (1994 supp. available).

Asselin, Thomas H. And Stout, L. Bruce. "Legal Exposure of the Design/Build Participants: The View of the General Contractor," *Construction Lawyer*, vol. 15, no. 3 (August 1995), pp. 8–24.

Case, Linda W. *Design/Build for Remodelers, Custom Builders, and Architects.* Washington, D.C.: Home Builder Press, National Association of Home Builders, 1989. 149 pp.

____________. *Remodelers Business Basics.* Washington, D.C.: National Association of Home Builders, 1989. 232 pp.

Citizen's Guide to Radon: What It Is And What To Do About It, A. 2nd ed. Washington, D.C.: U.S. Environmental Protection Agency, U.S. Department of Health and Human Services, and U.S. Public Health Service, 1992.

Collier, K. *Construction Contracts.* Englewood Cliffs, N.J.: Prentice Hall, 1987. 320 pp.

Householder, Jerry, and John C. Mouton. *Estimating Home Builders.* Washington, D.C.: Home Builder Press, National Association of Home Builders, 1992. 181 pp.

"Lead-Based Paint: Interim Guidelines for Hazard Identification and Abatement in Public and Indian Housing." *Federal Register* 55, No. 75 (April 18, 1990, and Sept. 28, 1990) 55FR39873.

NAHB Remodelors® Council and National Research Foundation. *Quality Standards for the Professional Remodeler.* Washington, D.C.: Home Builder Press, National Association of Home Builders, 1991. 81 pp.

Primer for Builders ... New Home Warranties and the Magnuson-Moss Warranty Act, A. Washington, D.C.: National Association of Home Builders, 1978. Available from the NAHB Consumer Affairs Department.

Schemm, Neil. "Subcontract Forms from the Subcontractor's Perspective." *Practical Real Estate Lawyer*, Sept. 1986. (Philadelphia: American Law Institute and Committee on Continuing Professional Education, American Bar Association.)

Smith, Carol. *The Positive Walk-Through.* Washington, D.C.: National Association of Home Builders, 1990. 139 pp.

____________. *Warranty Service for Builders and Remodelers.* Washington, D.C.: Home Builder Press, 1991. 191 pp.

Stein, Steven. *Construction Law*, vol. 1. New York: Mathew Bender, 1995.

Stokes, McNeil. *Construction Law in Contractors' Language*, 2nd ed. N.Y.: McGraw-Hill, 1990. 340 pp.

What Remodelers Need to Know and Do About Lead. Washington, D.C.: Technology and Codes Department and Remodelors® Council, National Association of Home Builders, 1992. 15 pp.

Whitney, Christopher C. "An Evolving Perspective on Design/Build Construction: A View from the Courthouse," *Construction Lawyer.* vol. 15, no. 2 (April 1995). pp. 1–99.

Order These Other Products from Home Builder Bookstore

Accounting and Financial Management for Builders

Emma Shinn

Explains how to design an accounting system and choose an accountant, defines terminology and procedures, and explains planning and analysis. Includes the NAHB Chart of Accounts and job cost control subsidiary ledger accounts.

Cost Control for Builders, Remodelers, and Developers

Jerry Householder

Describes how to plan, institute, and operate a cost control system that can help you improve profits. This system covers such crucial elements as comparing costs to estimates, analyzing variances, reducing overhead, keeping records, and planning future jobs. Explains how to start simply and gradually increase the level of detail. Provides cost control tools you can use right away, such as checklists, forms, and other examples.

Design/Build for Remodelers, Custom Builders, and Architects

Linda W. Case

Covers marketing and selling the service, handling contracts, and creating an efficient flow-through technique for completing jobs from start to finish. Interviews with practicing design-build remodelers, custom builders, and architects provide real-life examples.

The Developers Guide to Endangered Species Regulation

Presents a brief overview of the Endangered Species Act and land use regulation. Provides detailed treatments of the listing process, prohibitions against "takings" of species or habitats, Section 7 consultations, planning habitat conservation, enforcement, and related federal and state programs. Each chapter stands alone as a reference on a different aspect of the Act and it's related regulations. Thorough citations guide you or your attorney to additional information on particular topics.

How to Hire and Supervise Subcontractors

Bob R. Whitten

Shows remodelers and builders how to work with subcontractors not against them, find subcontractors who can do the job, communicate effectively with them, improve scheduling, maintain quality and cost controls, and more.

Quality Standards for the Professional Remodeler

Many remodelers incorporate into their contracts (by reference) this manual of 190 acceptable professional construction practices and standards for the remodeling industry.

Sample Letters and Memos for Builders, Developers, and Remodelers: Business Writing for Everyday Use

John A. Kilpatrick

Includes diskette with more than 90 sample letters and memos you can use for business management, marketing, sales, customer service, managing employees, and more. Load these samples right into WordPerfect (formatted for 5.1) and start using them right away. Shows remodelers, builders, and developers how to (a) use effective communication to improve professional image and (b) create a valuable paper trail in dealing with prospects, customers, suppliers, subcontractors, lenders, employees, and others.

Understanding House Construction

Walks the reader through the basics of new home construction from ground breaking to final inspection. Photos, text, and charts illustrate each step of the building process. Includes comprehensive glossary.

To place an order or for more information write or call—

Home Builder Bookstore Orders
1201 15th Street, NW
Washington, DC 20005-2800
(800) 223-2665